Component 1 Global Geographical Issues

This is assessed by Paper 1 in the exam. There are three sections, each on different topics. All questions are compulsory.

- **Section A**: Topic 1 *Hazardous Earth*, with questions on the *Global climate system*, *Climate change*, *Extreme weather* (e.g. tropical cyclones) and *Tectonic hazards*.
- **Section B**: Topic 2 *Development dynamics*, with questions on *Global development*, and a case study of one of the world's emerging countries.
- **Section C**: Topic 3 *Challenges of an urbanising world*, with questions on *Rapid urbanisation* and *Global urban trends*, and a case study of one of the world's megacities in either a developing or an emerging country.

In addition, questions will assess your geographical skills (such as interpreting statistics, maps, diagrams) within each topic.

Memory jogger for Paper 1!

- My case study of an emerging country was of

  _____.

- My case study of a developing or an emerging country was of

  _____,

  which is a developing/ emerging country (*delete one*).

- My case study of a megacity was _____
  (name of city) which is in

  _____
  (name of country).

Component 2 UK Geographical Issues

This is assessed by Paper 2 in the exam. There are four sections, each on a different topic, with a choice of fieldwork questions.

- **Section A** includes Topic 4 *The UK's evolving physical landscape* with questions on the UK's physical landscape, *Coastal change and conflict*, and *River processes and pressures*.
- **Section B** includes Topic 5 *The UK's evolving human landscape* with questions on the *UK's changing population* and a case study of one major UK city.
- **Sections C1 and C2** includes Topic 6 *Geographical investigations* with questions on fieldwork in Section C1 on **either** *Coastal change and conflict* **or** *River processes and pressures*, and in Section C2 on **either** *Dynamic urban areas* or *Changing rural areas*.

Like Paper 1, questions will assess your skills within each topic.

Memory jogger for Paper 2!

- My case study of a major UK city was of

  _____.

- My physical fieldwork was on rivers/coasts (*delete one*) and we collected data on
  _____ at

  _____
  (*name of place*).

- My human fieldwork was on cities/rural areas (*delete one*) and we collected data on
  _____ at

  _____
  (*name of place*).

Component 3 Making Geographical Decisions

This is assessed by Paper 3 in the examination. It consists of an unseen Resource Booklet with the following sections:

- **Section A**: Topic 7 *People and the biosphere*
- **Section B**: Topic 8 *Forests under threat*
- **Section C**: Topic 9 *Consuming energy resources*
- **Section D**: A geographical decision-making question

All three exam papers are quite different from each other.

Format of Paper 1

- **Time**: 1 hour 30 minutes.
- **Worth**: 94 marks in total – 90 marks on the three topics you've learned, and another 4 for spelling, punctuation, grammar and use of specialist geographical terminology (SPaG) which is assessed on one 8-mark question in Section B.
- **Counts for**: 37.5% of your final grade.
- **Number of sections**: three, assessing the topics described in Component 1 on page 5.

You must answer all questions as follows:

- **Section A**: *Hazardous Earth* – questions on the *Global climate system*, *Climate change*, *Extreme weather* (e.g. tropical cyclones) and *Tectonic hazards*. This section has 30 marks.
- **Section B**: *Development dynamics* – questions on *Global development*, and a case study of **one** of the world's emerging countries. This section has 30 marks.
- **Section C**: *Challenges of an urbanising world* – questions on *Rapid urbanisation* and *Global urban trends*, and a case study of one of the world's megacities in **either** a developing **or** an emerging country. This section has 30 marks.

The 8-mark question in Section B in Paper 1 carries an additional 4 marks for SPaG.

Any resources that you need to answer the questions are included – there is no separate Resource Booklet for Paper 1.

Format of Paper 2

- **Time**: 1 hour 30 minutes.
- **Worth**: 94 marks in total – 90 marks on topics you've learned, and another 4 for SPaG, which is assessed on one question in Section B.
- **Counts for**: 37.5% of your final grade.
- **Number of sections**: four, assessing the topics described in Component 2 on page 5.

You must answer:

- **all** parts of **Section A**: *The UK's evolving physical landscape* – questions on the *UK's physical landscape*, *Coastal change and conflict*, and *River processes and pressures*. This section has 27 marks.
- **all** parts of **Section B**: *The UK's evolving human landscape* – questions on the *UK's changing population* and a case study of one major UK city. This section has 27 marks plus 4 marks for SPaG.
- **two** questions on fieldwork in **Section C**. One must be from Section C1 on **either** *Coastal change and conflict* **or** *River processes and pressures*, and the second from Section C2 on **either** *Dynamic urban areas* **or** *Changing rural areas*. Each of these has 18 marks.

Any resources you need to answer the questions are included – there is no separate Resource Booklet for Paper 2.

GCSE 9-1

geography

EDEXCEL B

SECOND EDITION

Exam Practice

SERIES EDITOR
Bob Digby   Nicholas Rowles   Kate Stockings

OXFORD
UNIVERSITY PRESS

OXFORD
UNIVERSITY PRESS

Great Clarendon Street, Oxford, OX2 6DP, United Kingdom

Oxford University Press is a department of the University of Oxford. It furthers the University's objective of excellence in research, scholarship, and education by publishing worldwide. Oxford is a registered trade mark of Oxford University Press in the UK and in certain other countries

Series editor: Bob Digby

Authors: Bob Digby, Nicholas Rowles, Kate Stockings

The moral rights of the authors have been asserted

Database right of Oxford University Press (maker) 2024

First published in 2019

Second edition 2024

British Library Cataloguing in Publication Data
Data available

ISBN 978-138-202915-5

10 9 8 7 6 5 4 3 2 1

The manufacturing process conforms to the environmental regulations of the country of origin.

Printed in Great Britain by Ashford Colour Press Ltd, Gosport.

Acknowledgements

The publisher and authors would like to thank the following for permission to use photographs and other copyright material:

Photos: p1: Revenant / Shutterstock; **p16:** Ann Rayworth / Alamy Stock Photo; **p17:** Crown copyright / Ordnance Survey; **p23:** Harvepino / Shutterstock; **p27:** Bob Digby; **p34:** imagegallery2 / Alamy Stock Photo; **p36:** imagegallery2 / Alamy Stock Photo; **p47:** Stuart Kelly / Alamy Stock Photo; **p52:** Davidovich Mikhail / Alamy Stock Photo; **p98:** GavidD / iStock; **p99:** Mika Schick / Alamy Stock Photo; **p107:** The Photolibrary Wales / Alamy Stock Photo; **p108:** Andrew Stacey / www.stacey.peak-media.co.uk; **p114:** Chris Green / Shutterstock; **p140:** Robert Thorley / Alamy Stock Photo; **p144:** Associated Press / Alamy Stock Photo; **p145:** Jake Lyell / Alamy Stock Photo; **p146:** Georg Gerster / Panos.

Artwork by Q2A Media, Aptara Inc., Kamae Design, Barking Dog Art, Lovell Johns, and Mike Phillips.

Ordnance Survey (OS) is the national mapping agency for Great Britain, and a world-leading geospatial data and technology organisation. As a reliable partner to government, business and citizens across Britain and the world, OS helps its customers in virtually all sectors improve quality of life.

Although we have made every effort to trace and contact all copyright holders before publication this has not been possible in all cases. If notified, the publisher will rectify any errors or omissions at the earliest opportunity.

Contents

Mark schemes for the Exam Practice Papers are available on the Oxford Secondary Geography website: **www.oxfordsecondary.co.uk/geog-edexcelb-answers**

Please note: The Practice Paper exam-style questions and mark schemes have not been written or approved by Edexcel. The answers and commentaries provided represent one interpretation only and other solutions may be appropriate.

If you want to be successful in your exams, then you need to know how you will be examined, what kinds of questions you will come up against in the exam, how to use what you know, and what you will get marks for. That's where this book can help.

How to use this book

This book contains the following features to help you prepare for exams for the Pearson Edexcel GCSE (9–1) Geography B specification. It is written to work alongside two other OUP publications to support your learning:

- *GCSE (9–1) Geography Edexcel B Student Book*
- *GCSE (9–1) Geography Edexcel B Revision Guide.*

Introduction (pages 4–12)

This section contains details about:

- the exam papers you'll be taking, and what you need to revise for each exam paper
- how exam papers are marked, and how to aim for the highest grades.

On your marks! (pages 13–91)

This section contains guidance about how to answer exam questions, varying from short answers (1–3 marks) to those requiring extended writing (4, 8 and 12 marks). There is space for you to write and assess exam answers so you learn how to write good quality answers. There are answers to these questions on pages 148–155.

Exam Practice Papers (pages 92–147)

This section contains one full set of exam papers. This is written to match the style of those you'll meet in the Pearson Edexcel GCSE (9–1) Geography B exam. This set contains:

- Exam papers which assess your knowledge and understanding of the course, together with fieldwork in Paper 2. You'll have two days of fieldwork during the course to help answer this part of the exam – one day physical geography and one day human geography.
- Exam Paper 3 which assesses a decision-making exercise (usually called a DME). This assesses your knowledge, understanding and skills in interpreting an unseen Resource Booklet. During the exam, you'll use the booklet to understand the geographical issues on which it is based, and your geographical skills in making sense of it.

Each exam paper has answers on the Oxford Secondary Geography website (www.oxfordsecondary.co.uk/geog-edexcelb-answers) for you to refer to.

The Pearson Edexcel GCSE (9–1) Geography B specification consists of three components. Each component contains topics. Each component is assessed by an exam paper (Papers 1, 2 and 3) with sections for different topics, as shown on the next page.

Format of Paper 3

- **Time**: 1 hour 30 minutes.
- **Worth**: 64 marks in total – 60 for geographical questions, and 4 marks for SPaG which are assessed in the final question in Section D.
- **Counts for**: 25% of your final grade.
- **Number of sections**: four

You must answer all questions as follows:

- **Section A**: questions on *People and the biosphere*. This section has 7–8 marks.
- **Section B**: questions on *Forests under threat*. This section has 7–9 marks.
- **Section C**: questions on *Consuming energy resources*. This section has 31–33 marks.
- **Section D**: a geographical decision-making question. This question carries 12 marks plus 4 marks for SPaG.

Paper 3 has a separate Resource Booklet.

Question style

The first questions in each section are short and worth between 1 and 4 marks.

- These include a mix of multiple-choice, short answers, or calculations.
- There are resource materials (data, photos, cartoons, etc.) on which you'll be asked questions. These could include statistical skills, so remember you can use a calculator in each exam.
- You'll be expected to know what the resource materials are getting at from what you've learned.
- Detailed case study knowledge is only needed for the case studies in Papers 1 and 2 – though you can get marks for using examples.

All these questions are **point marked**.

Later questions in each section require extended writing, and are worth 8 marks. Paper 3 also contains one question of 12 marks. You need to have learned examples and case studies to answer these questions. Answers like this are marked using **levels** – from Level 1 (lowest) to Level 3 (highest) – see page 10. These questions are likely to be on those parts where you have been taught examples (e.g. on Paper 1 a tectonic hazard, a tropical cyclone event, or an emerging country).

One 8-mark question on each paper is also assessed for SPaG for 4 marks – making it worth 12 marks in total.

Answering questions properly is the key to success. When you first read an exam question, check out the **command word** – that is, the word that the examiner uses to tell you what to do. Figure 1 gives you the command words you can expect, and the number of marks you can expect for each command word.

Command word	Typical no. of marks	What the command word means	Example of a question
Identify/State/Name	1	Find (e.g. on a photo), or give a simple word or statement	Identify the landform shown in the photo *or* Name City *a* on map 2.
Define	1	Give a clear meaning	Define the term 'fertility rate'.
Calculate	1 or 2	Work out	Calculate the mean depth of the river shown in Figure *W*.
Label	1 or 2	Print the name of, or write, on a map or diagram	Label the two features A and B of the cliff in Figure *X*.
Draw	2 or 3	As in sketch or drawing a line	Draw a line to complete the graph in Figure *Y*.
Compare	3	Identify similarities or differences	*(referring to a graph)* Compare the rate of population growth in city *b* with city *c*.
Describe	2 or 3	Say what something is like; identify trends (e.g. on a graph)	Describe one feature on the photo shown in Figure *Z*.
Explain	2, 3 or 4	Give reasons why something happens	Explain the rapid growth of one named megacity you have studied.
Suggest	2, 3 or 4	In an unfamiliar situation (e.g. a photo or graph), explain how or why something might occur, giving a reason.	Suggest reasons for the increase shown in the graph.
Assess	8	Weigh up which is most/least important	Assess the impacts of a tropical storm upon a named country.
Evaluate	8	Make judgments about which is most or least effective	Evaluate the reliability of the conclusions in your fieldwork.
Justify	12	Give reasons why you support a particular decision or opinion	*(in Paper 3, last question)* Justify the reasons for your choice.

Figure 1 *Meaning of command words, marks available and examples*

Examiners have clear guidance about how to mark. They must mark fairly, so that the first candidate's exam paper in a pile is marked in exactly the same way as the last. You will be rewarded for what you know and can do; you won't lose marks for what you leave out. If your answer matches the best qualities in the mark scheme then you'll get full marks.

Questions that carry between 1 and 4 marks are **point marked**, and those carrying 8 marks or more are **level marked**. Be clear about what this means.

Understanding Assessment Objectives

Assessment Objectives (called AOs) are the things that examiners look for in marking your answers. There are four in GCSE Geography:

- AO1 – Knowledge recall
- AO2 – Understanding of concepts, places and environments
- AO3 – Applying ideas to situations, and making informed judgments
- AO4 – Geographical skills, which includes fieldwork, stats and maths skills

Command word	Assessment Objective	Example of a question assessing this AO
Identify/State/Name	AO1	Identify the landform shown in the photo *or* Name City *a* on Map 2.
Define	AO1	Define the term 'fertility rate'.
Calculate	AO4	Calculate the mean depth of the river shown in Figure *W*.
Label	AO1	Label features A and B of the cliff in Figure *X*.
Draw	AO4	Draw a line to complete the graph in Figure *Y*.
Compare	AO3	Compare the rate of population growth in city *b* with city *c*.
Describe	AO1 & 2	Describe the features of a river meander.
Explain	AO1 & 2	Explain the rapid growth of one megacity you have studied.
Suggest	AO3	Suggest reasons for the increase shown in the graph.
Assess (8 marks)	Paper 1: AO2 and AO3 (4 marks each) *or* Paper 2: AO3 and AO4 (4 marks each)	*(Paper 1)* Assess the impacts of a named tropical cyclone. *(Paper 2)* Assess the potential impacts of the regeneration scheme shown in Figures 1 and 2.
Evaluate (8 marks)	Paper 1: AO2 and AO3 (4 marks each) *or* Paper 2: AO3 and AO4 (4 marks each)	*(Paper 1)* Evaluate whether tectonic hazards have greater impacts on developed than on developing countries. *(Paper 2)* Evaluate the accuracy of the methods used in collecting data in your physical fieldwork.
Justify (12 marks)	AO2, AO3 and AO4 (4 marks each)	*(in Paper 3, last question)* Justify the reasons for your choice.

Figure 2 *Examples of the command words used for each AO and typical questions. 'Assess', 'Evaluate' and 'Justify' are the most challenging.*

Understanding the most demanding questions

The last three command words in Figure 2 are the most demanding. They

- assess AO3
- form the extended written questions
- carry the highest marks.

Figure 2 shows that questions often combine marks for AO3 with marks for other AOs. Examples where AO3 is used with AO2 include:

- *Assess the impacts of a named tropical cyclone (8 marks).* This involves you knowing and understanding impacts of tropical cyclones (that's AO2 – understanding) for 4 marks and then assessing how serious each impact is (so that's AO3 – application) for another 4 marks.

Sometimes AO3 is used with AO4. Examples include:

- *Evaluate the accuracy of the methods used in collecting data in your fieldwork (8 marks).* This involves you using your experience of fieldwork skills (AO4) for 4 marks and then judging how accurately these methods proved (so that's AO3 – application) for another 4 marks.

In Paper 3, three AOs are assessed in the final 12-mark question:

- *Justify the reasons for your choice (12 marks).* This involves using data in the Resource Booklet (AO4) for 4 marks, using what you have learned in the course (AO2) for another 4 marks, and then how you use these to make a judgment about which is the best option (that's AO3 – application) for a further 4 marks.

Note in each case that to get full marks, you must address each AO.

Level-marked questions

Longer questions worth 8 or 12 marks are marked using levelled mark schemes. Examiners mark answers based on these. Figure 4 shows an 8-mark scheme.

- There are three levels; Level 1 is the lowest and Level 3 the highest.
- Figure 3 is a detailed summary of what examiners look for.
- Note that it is not so much the number of points you make that matter, but the ways in which these are explained and extended.

 Tip

You can find more detailed guidance on answering 8- and 12-mark questions in the 'On your marks' section on pages 50–91.

Level	Marks	Descriptor
	0	No acceptable response.
1	1–3	• Limited or no explanation. • One or two points are simply described but not developed. • Most of the answer lacks detail or named examples. • Places are poorly located (e.g. 'in Africa'). • Few geographical terms or phrases. • Makes no judgment when asked to 'assess' or 'evaluate'.
2	4–6	• Some fairly clear explanation. • Two or three points are explained briefly with some development. • Examples are used, but vary in detail; places (e.g. countries) and impacts are named. • Writes clearly, using some geographical terms. • Makes some judgment when asked to 'assess' or 'evaluate'.
3	7–8	• Explains very clearly. • Makes detailed points, using extended explanations to develop the answer. • Detailed examples are used; specific places and impacts are named. • Well written, with full use of geographical terms. • Makes detailed judgments based on evidence when asked to 'assess' or 'evaluate'.

Figure 3 Level marking – a summary of what examiners look for

Using the command word 'Assess'

When they mark, examiners do not mark points, but instead read the answer as a whole, and judge it against the qualities shown in Figure 4 – a detailed level-based mark scheme for all 8-mark questions which assess AO2 and AO3. For 8 marks you need about three extended and exemplified points. Level 3 is reserved for candidates who 'assess' as the command word tells them to do.

Level	Marks	Descriptor
	0	No acceptable response.
1	1–3	• Limited understanding of concepts and links between places, environments and processes. (AO2) • Some application and understanding, though links may be flawed. (AO3) • An imbalanced or incomplete argument showing little understanding. Judgments supported by limited evidence. (AO3)
2	4–6	• Some understanding of concepts and links between places, environments and processes. (AO2) • Generally applies understanding to deconstruct information and provide some logical links between concepts. (AO2) • An imbalanced argument – mostly coherent explanations, leading to generalised judgments supported by some evidence. (AO3)
3	7–8	• Accurate understanding of concepts and links between places, environments and processes. (AO2) • Applies understanding to deconstruct information and provide logical links between concepts. (AO3) • A balanced, well-developed argument with coherent relevant explanations, leading to judgments supported by detailed evidence. (AO3)

Figure 4 *A mark scheme using levels of response*

How to 'assess' and produce a Level 3 answer

Figure 5 shows a Level 3 answer to the question *'Assess the economic effects of a named tropical storm'. (8 marks)*. It is worth the full 8 marks.

The storm is named. Always do this – you could limit yourself to Level 1 or 2 if you don't. Notice the phrase *'most serious'* – it is evidence the candidate is assessing this impact.

Credit is given for the detailed explanation of the economic effects of a storm – *'damaged buildings'*, *'cost insurance companies and governments millions'*. This makes the answer Level 3 because of the detail. The candidate is also assessing the storm as the 'worst known'.

'In Cyclone Aila which affected Bangladesh in 2009 the economic effects were enormous, and were most serious among the poor. The cyclone brought some of the worst known stormy winds and floods. This damaged buildings, which cost insurance companies and governments millions. More storms also caused erosion of flood defences, which flooded villages and farmland costing huge amounts to replace and repair, and destroyed crops. For many farmers and families, this meant loss of homes and crops, making their poverty worse and forcing some to leave the land and move to Dhaka, the capital, for work.'

'erosion of flood defences' is an economic impact. The phrase *'costing huge amounts'* is a judgment of its seriousness, making this answer Level 3, and showing that the candidate is assessing.

'loss of homes and crops' is evidence of another economic impact. The candidate assesses the seriousness of loss of crops – it makes poverty worse and forces people to leave the land. This judgment answers the command word 'Assess'.

Figure 5 *An example of a Level 3 answer*

Always make sure you answer the question that is set!

- Good answers are usually focused – without straying off the point.
- Good answers are more likely if you unpick the question, as shown below.

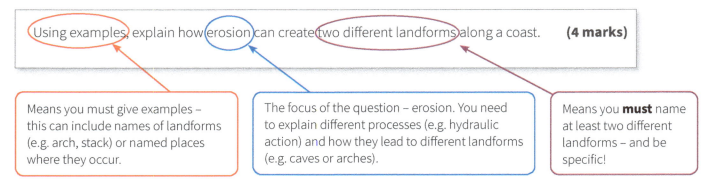

Using examples, explain how erosion can create two different landforms along a coast. **(4 marks)**

Means you must give examples – this can include names of landforms (e.g. arch, stack) or named places where they occur.

The focus of the question – erosion. You need to explain different processes (e.g. hydraulic action) and how they lead to different landforms (e.g. caves or arches).

Means you **must** name at least two different landforms – and be specific!

This question is about explaining processes that lead to different landforms such as arches. It's important that you explain why these processes happen – don't just describe.

Using case studies

Candidates worry about case studies – how to learn them and write good answers. The following question requires examples from your case study of an emerging country.

'Assess the impacts of economic growth on the population of a named emerging country.' **(8 marks)**

This question could be answered using examples from India (student book section 2.11).

One way to plan case studies is to use a spider diagram.

1 Wealth generated by economy means money for healthcare

2 Improved healthcare in cities so child and maternal mortality are reduced

Impacts of economic growth on India's population

3 Women may have more freedoms in cities, so marry later and have fewer children

4 Falling birth rate, especially in cities, so population growth slows down

From this, build up notes, e.g.:

- from box **1**, draw two more 'legs' to show how a growing economy leads to **a** investment in healthcare and **b** hospitals offering better treatment
- from box **2**, draw three more 'legs' to explain how **a** more women give birth in hospitals, **b** receive pre-natal care, and **c** receive treatment if anything goes wrong
- from box **3**, draw three more 'legs' to explain how **a** caste traditions may be lost, **b** women may enjoy careers, and **c** women have freer choice of marriage partners
- from box **4**, draw two more 'legs' to show **a** people may have fewer children in cities and **b** because housing and education costs may be high.

On your marks!

This section focuses on how you can maximise your chances of high marks in the examination papers. The contents of this section are shown below, and are organised by the number of marks per question.

On your marks!

1 Mopping up the 1-mark questions

1.1 Multiple-choice questions

The phrase 'multiple choice' in exam questions means exactly that – you are given a situation in which (usually) four possible answers are given, but only one is correct. That's your job – to find the correct answer! These questions are normally worth 1 mark, and you have a choice of one from four possible answers.

What are the common mistakes?

Even top students often perform poorly on multiple-choice questions, throwing away important marks. Don't rush! They may only be worth 1 mark but this doesn't mean they're easy! Common mistakes include giving too many answers.

Tip

Find the correct answer by crossing out the choices you **know** are wrong. That leaves one or two possible answers. This forces you to take your time choosing, making it more likely you'll get the correct answer.

> **Worked example**
>
> Study **Figure 1**, a graph showing variations in average temperature in Australia (1910–2018).
>
>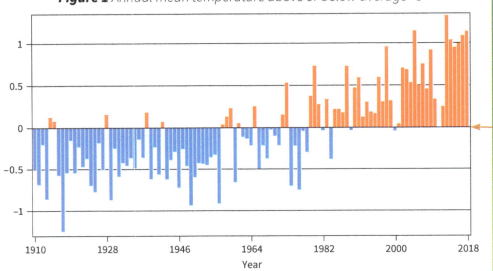
>
> *Figure 1 Annual mean temperature above or below average °C*
>
> Using **Figure 1**, identify which **one** of the following statements is true.
>
> ☒ **A** ~~Australia's temperatures were below average every year before 1946.~~
>
> ☒ **B** ~~Australia's temperatures showed a steady increase between 1982 and 2018.~~
>
> ☒ **C** Australia's temperatures have been above the 1961–1990 average every year since 2001.
>
> ☒ **D** ~~Since 2000, Australia's temperatures have exceeded 0.5°C above the 1961–1990 average.~~
>
> **(1 mark)**

This is the long-term average temperature. The blue bars show years where the average temperature was below the long-term average. The red bars show where the average temperature was above the long-term average.

Take time to look closely at both axes to make sure you understand the scales. Think about what's happening over time – what's the story?

Carefully work your way through the four options, looking closely at the graph. Try to identify the options that are definitely incorrect and draw a line through them. Take your time to make the correct decision. By elimination, you should be left with the correct answer.

Now try these!

1   Study **Figure 1**, which shows the tectonic plates in Iceland.

Figure 1

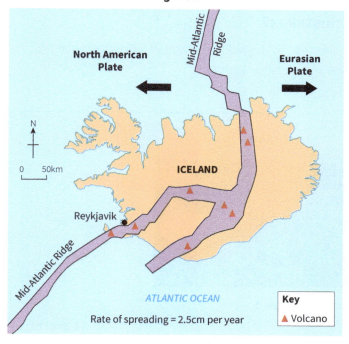

Using **Figure 1**, identify how long it will take the plates to spread by 100 m.

☒ **A**   400 years

☒ **B**   4000 years

☒ **C**   2500 years

☒ **D**   40 years

(1 mark)

2   Identify which term is best defined by the phrase 'the unplanned growth of urban areas into the surrounding countryside'.

☒ **A**   Urbanisation

☒ **B**   Urban regeneration

☒ **C**   Urban greening

☒ **D**   Urban sprawl

(1 mark)

3   Identify which of the following best describes how sedimentary rocks are formed.

☒ **A**   Formed from molten rock beneath the Earth's surface

☒ **B**   Formed from molten rock upon the Earth's surface

☒ **C**   Formed when earth movements occur

☒ **D**   Formed from material laid down by rivers, moving ice or the sea

(1 mark)

1.2 Skills questions

Skills are an important part of your Geography studies. They include interpreting resources such as maps, diagrams, graphs or tables of data. Your exams will feature a number of questions designed to test your ability to use them.

What are the common mistakes?

Skills-based questions vary in difficulty. In the worked example below, you need to spot evidence in order to answer the question properly.

Worked example

Figure 1

1   Identify the landform at **A** in **Figure 1**.
    Stack
    (1 mark)

2   Identify one piece of evidence from **Figure 1** that this feature once formed part of a coastal arch.
    There is debris between the stack and the mainland, which is the remains of an arch.
    (1 mark)

Question 1 tests your knowledge. It just needs a one-word answer that names the landform.

Question 2 needs some interpretation. You need to know what a coastal arch is. If you know that landform A is a stack, you'll know it formed when an arch collapsed. The rocks show the rubble from when it collapsed!

Now try these!

1   **Figure 1** shows how the ecosystem changes depending on the height above sea level.

Figure 1

Height above sea level	Ecosystem
0–900 m	Tropical rainforest
900–1800 m	Temperate forest

(a) Using **Figure 1**, draw lines to complete the line graph for tropical rainforest and temperate forest in **Figure 2**.

(1 mark)

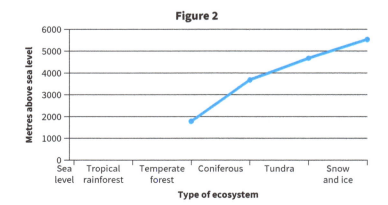

(b) Identify the data from **Figure 2** to complete the following sentence.

(1 mark)

Coniferous forest exists between _____ metres and _____ metres above sea level.

(c) Identify the height at which tundra gives way to snow and ice.

(1 mark)

_____

2 Study **Figure 3**, a 1:25 000 Ordnance Survey map of the area around the Eden Project in Cornwall.

Figure 3

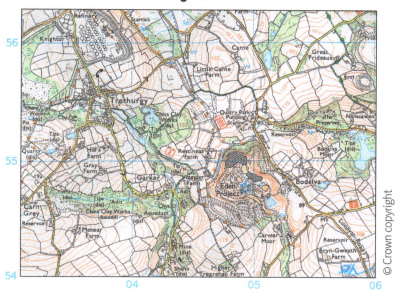

© Crown copyright

(a) Using **Figure 3**, identify the four-figure grid reference for the grid square containing the Eden Project.

⊠ **A** 0355

⊠ **B** 0454

⊠ **C** 0556

⊠ **D** 0354

(1 mark)

(b) Identify in which direction Trethurgy (0355) lies from the Eden Project.

(1 mark)

_____

(c) Identify one piece of evidence from **Figure 3** that the china clay industry was once important in the area of the map extract.

(1 mark)

_____

1.3 Knowledge questions

'Knowledge' questions are questions that test you directly on what you know. Expect to see questions that ask you to define key terms, or to select one of many terms that you have learned about.

What are the common mistakes?

These questions vary in difficulty – don't expect that they will be easy just because they are 'only' worth 1 mark.

- Basically, it comes down to whether you know the answer or not!
- Almost always, there is more than one way to give the answer, as the worked example below shows.

> **Worked example**
>
> Define the term 'abiotic'.
>
> **(1 mark)**
>
> *Things that don't include plants and animals in an ecosystem, such as water.*
>
> Only one answer line is given, so only a brief phrase is expected. Other possible answers to this question could include:
>
> - 'non-living parts of an ecosystem' (or a food web)
> - 'non-biotic', 'not biotic' or 'not biological'
> - giving an example, such as 'rocks', 'minerals', 'water', 'soil'.

> **Tip**
>
> When revising, make a list of all the key words and phrases you need to know from the specification.
>
> When you answer a question, use the number of lines to guide how much you think you should write.
>
> - If one line is shown, only a word or brief phrase is expected.
> - If two lines are shown, a longer phrase or sentence is expected.

> **Now try these!**
>
> Try these questions with one line answer space. Time yourself – 4 minutes!
>
> 1   Name **one** characteristic of a sedimentary rock.
>
> **(1 mark)**
>
> _____
>
> 2   Name **one** type of mechanical weathering that might have an impact on UK landscapes.
>
> **(1 mark)**
>
> _____
>
> 3   Farming is one example of a human activity affecting the landscape.
>   Name **one** other example of a human activity that affects the landscape.
>
> **(1 mark)**
>
> _____
>
> 4   State **one** natural cause of climate change.
>
> **(1 mark)**
>
> _____

Now try these!

Try these questions with two lines of answer space. A sentence is needed.

5   State **one** piece of evidence that can be used to show natural climate change in the past.

(1 mark)

_____

_____

6   Define the term 'counter-urbanisation'.

(1 mark)

_____

_____

7   State the meaning of the initials 'HDI'.

(1 mark)

_____

_____

8   State **one** feature of a meander along a river.

(1 mark)

_____

_____

9   State **one** primary effect of a tectonic hazard.

(1 mark)

_____

_____

10 Define the term 'renewable energy'.

(1 mark)

_____

_____

11 Define the term 'erosion'.

(1 mark)

_____

_____

On your marks!
2 Maxing out the 2-mark questions

2.1 How the 2-mark questions work

Sometimes examiners want you to describe or explain something in a little more detail. This means that you need to develop your answers with more detail in order to get the marks.

Look at this question:

> Rainforests are ecosystems. State **two** ways in which humans can protect ecosystems. **(2 marks)**

There are **2** marks for this question, so you must suggest **two** ways of protecting ecosystems. You get 1 mark for each way or point that you give. The mark scheme tells examiners which ways they can mark as correct. It's just like the 1-mark questions, except you have to make two points!

Developing your answers for 2 marks

Some questions ask you to describe or explain one thing for 2 marks. You need to know how to turn a 1-mark answer into a 2-mark answer. This means **developing** a point. Look at the following question:

> Explain **one** possible economic impact of climate change. **(2 marks)**

In this case, it is not enough to just *name* one economic impact – this would earn just 1 mark. To earn 2 marks your answer must be either:

- **developed** (i.e. explained with extra detail), or
- **exemplified** (i.e. giving an example of what you are explaining).

'Describe' and 'Explain'

Make sure that you know the difference between '**Describe**' and '**Explain**'.

'**Describe**' means you need to say what is there, what you see, or what something is like.

'**Explain**' means you need to say *how* or *why* something occurs.

> **Tip**
>
> To develop an answer, extend it using one of the following phrases:
>
> - 'so that…'
> - 'which means that…'
> - 'which leads to…'
> - 'therefore…'
> - 'for example…'
> - 'because…'

Worked example

1 Explain **one** possible economic impact of climate change. **(2 marks)**

   Hotter summers might cause crops to die ✓ and farmers would lose their income. ✓

   - For 2-mark questions, you should extend the answer with what happens as a result, or with an example of how something occurs.

2 Explain **one** natural cause of climate change. **(2 marks)**

   One cause might be when a volcano erupts ✓ because the ash caused by the explosion blocks out the sun's rays. ✓

> One economic impact is shown for 1 mark.

> An additional detail is given for a second mark.

> One economic impact is shown for 1 mark.

> The effect of the explosion is given for a second mark.

Now try these!

 Tip

'**Describe**' means to say what something is like, or what exists. So for question 1, think what life in a large developing world city is like, and simply say what it's like. For question 2, say what a tropical cyclone consists of. If you drew a diagram of it, what would you label?

1   Describe **one** feature of the quality of life in a major city in an emerging country.

(2 marks)

_____

_____

_____

_____

2   Describe **one** feature of the structure of a tropical cyclone.

(2 marks)

_____

_____

_____

_____

Tip

'**Explain**' means to give reasons why something is as it is or how it happens. So, for question 3, name the way in which economic development is measured and say *how* it does this. For question 4, say *why* a country might want investment, and why there might be benefits.

3   Explain **one** way in which a country's economic development can be measured.

(2 marks)

_____

_____

_____

4   Explain **one** way in which high levels of investment could benefit an emerging country.

(2 marks)

_____

_____

_____

2.2 Skills questions

Many questions will ask you to look at a figure, such as a graph, map, photo or table, and then describe it. Figures like this usually contain data (i.e. numbers). These questions are testing your skills and ability to get information from the figure.

In questions like this, you could be asked to describe:

*   a change over time – known as a **trend** or **variation**
*   differences between places – a **pattern** or **distribution**.

You probably won't have seen the graph before, so you won't be asked to explain or give reasons in these questions.

Sometimes, the command word 'suggest' is used, because you can't be expected to know a particular example. But you can use intelligent guesswork to work out how or why something might happen.

Tip

If you are asked to describe a graph, look for these things:

*   **The trend**. Is the data increasing or decreasing (or does it do both)?
*   **The x-axis**. What are the years (or other units of time) over which the increase or decrease occurred?
*   **The y-axis**. Give some figures to illustrate your answer. What was the value at the lowest point? Or at the highest point?
*   **The range**. Calculate the difference between the highest and lowest points.

Activity 1

Figure 1 shows mean global sea level to 2020, and a range of predictions about future changes by 2100.

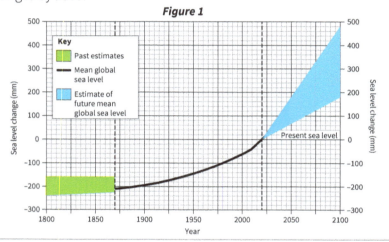

Figure 1

Describe the estimated future mean global sea level. **(2 marks)**

Now look at the answers below. Decide how many marks each answer earns, and give a reason. Write your judgment on the lines provided.

Answer 1: Sea level is predicted to rise.

Answer 2: Sea level is predicted to rise. The predictions have a large range.

Answer 3: Sea levels are changing because the ice caps are melting.

Answer 4: Predicted increases have a wide range, from +180 to +480mm by 2100.

Now try these!

1 **Figure 1** shows a satellite photo of Hurricane Sandy off the coast of Florida, USA in 2012.

Figure 1

Using **Figure 1**, describe **one** feature of Hurricane Sandy.

(2 marks)

_____

_____

_____

_____

Figure 2

| Journey time in minutes | Percentage of all journeys to work | |
	Megacity	Rest of the country
1–15	50	82
16–30	20	8
31–45	15	6
46–60	5	4
Over 60	10	0

2 **Figure 2** shows the percentage of journey times to work for workers in a megacity in an emerging country compared with the rest of the country.

Describe **one** difference between journey times to work for people in the megacity and journey times for people in the rest of the country.

(2 marks)

_____

_____

_____

_____

2.3 Knowledge questions

Questions that test your knowledge and understanding often ask for some detail.

- They need more detail than answers to 1-mark questions (see pages 14–19).
- This means that you need to develop your answers in the ways shown on pages 20–23.
- You would usually be asked to describe a feature or process, or suggest or explain a reason why something happens.

Tip

Questions testing your knowledge usually have the following command words:

- **'Suggest'** e.g. 'Suggest one reason for the trend shown in the graph.'
- **'Describe'**, e.g. 'Describe one landform in a river's upper course.'
- **'Explain'**, e.g. 'Explain one reason for the rapid growth in street dweller numbers in megacities in emerging countries.'

Worked example

Using **Figure 1**, explain **one** effect of global atmospheric circulation.

(2 marks)

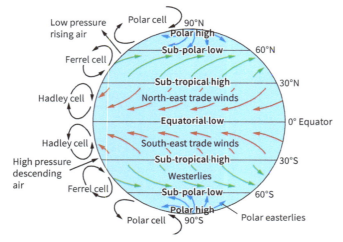
Figure 1 Global atmospheric circulation

Answer 1: Global atmospheric circulation carries heat and moisture to other places ✓, e.g. the UK gets warmth from south-westerly winds from warm sub-tropical areas. ✓

Answer 2: Global atmospheric circulation forms currents of warm or cool air. ✓

Answer 3: Global atmospheric circulation forms winds, ✓ which heat up the north Atlantic and make it warmer than it would be without any current. ✓

Examiner feedback

Answer 1 is worth 2 marks. The candidate correctly says that global atmospheric circulation carries heat and moisture to other places. This is then **developed** by explaining the effect of south-westerly winds in the UK.

Examiner feedback

Answer 2 is only worth 1 mark. The candidate has mentioned the currents of warm or cool air, but not what effect these have.

Examiner feedback

Answer 3 is worth 2 marks. The candidate says what global atmospheric circulation does for 1 mark. Then this is **developed** with the example of how this heats the north Atlantic, gaining the second mark.

Now try these!

1 Describe **one** way in which a region affected by volcanic eruptions can prepare for this hazard.

> **Tip**
>
> For a reminder of how to 'describe', see page 21.

(2 marks)

_____

_____

_____

_____

2 Describe **one** way in which one named major UK city is trying to make transport more sustainable.

(2 marks)

_____

_____

_____

_____

3 Explain **one** feature of a constructive plate margin.

> **Tip**
>
> For a reminder of how to 'explain', see page 21.

(2 marks)

_____

_____

_____

_____

4 Explain **one** reason why developing or emerging countries often suffer worse impacts of tropical storms than developed countries.

(2 marks)

_____

_____

_____

_____

_____

Maxing out the 2-mark questions

2.4 'Describe' and 'explain' questions

Students often mix up 'describe' and 'explain'.

- **Describe** means that you have to write an account of the main features of something (e.g. the main features of urban regeneration) or the steps in a process (e.g. how climate change occurs). Most 'describe' questions are worth 2 or 3 marks and so need to be developed. However, you don't have to give any reasons.

- **Explain** means that you have to give reasons for how or why something occurs (e.g. the causes of urban regeneration). You often have to give an example of something. You could even be asked to draw and label diagrams to support your explanation.

Tip

Make sure you are clear about the difference between 'describe' and 'explain':

- **Describe:** tell it like it is, but don't give reasons why it is! Give information that paints a picture.
- **Explain:** give reasons for how or why something happened.

Worked example

Think about this exam question. Look at the two answers below and make sure you understand why each one receives the marks it does.

Describe **one** feature of an urban regeneration project. **(2 marks)**

'several sports venues' is a descriptive point, so earns the candidate 1 mark.

Answer 1:

East London's Queen Elizabeth Olympic Park has several sports venues which were designed to improve the economy of the area.

'which were designed to improve the economy of the area' is a reason for building them – so it's an explanation, not a description. This means this answer only receives 1 mark.

Answer 2:

East London's Queen Elizabeth Olympic Park has several sports venues, such as the London stadium.

The phrase 'such as the London stadium' is also descriptive – the candidate has used an example to develop the answer, which gets the second mark.

'several sports venues' is a descriptive point, so earns the candidate 1 mark.

Tip

The word 'because' makes connections between cause and effect, so should be used in 'explain' questions but not in 'describe' questions.

Now try these!

1 Describe **one** method used by regions affected by tropical cyclones to prepare for this hazard.

(2 marks)

_____

_____

_____

_____

2 Explain **one** way in which a region affected by tropical cyclones can prepare for this hazard.

(2 marks)

_____

_____

_____

_____

> Did you spot the command word? Check that you have **described** for question 1, and **given reasons** for question 2.

3 Using **Figure 1**, describe **one** way in which trees adapt to the tropical rainforest.

(2 marks)

Figure 1

_____

_____

_____

_____

4 Explain **one** way in which trees adapt to the tropical rainforest environment.

(2 marks)

_____

_____

_____

> Check that you have **described** for question 3, and **given reasons** for question 4.

On your marks!

3 Tackling the 3-mark questions

3.1 Chains of reasoning

Sometimes, examiners want a little bit more from you than just a single developed answer. This is often the case when there is quite a detailed process involved, or a more complicated explanation. In these cases, examiners set a 3-mark question. The examiners are expecting you to give a 'chain of reasoning' – a sequence of statements, one following on from the other.

What are the common mistakes?

Candidates generally average only 1 mark on 3-mark questions. As with 2-mark questions, the mistake of saying just one thing without any development is common.

> **Worked example**
>
> Explain **one** way in which ecosystems such as rainforests can be protected.
>
> **(3 marks)**
>
> With 3-mark questions, it is not enough to just name one way in which ecosystems can be protected. To earn 3 marks, you must extend the point (i.e. explain in more detail) twice, as in this example:
>
> Ecosystems such as rainforests can be protected by making it illegal to carry out logging ✓ so that forest habitats are protected for animals ✓ which maintains the biodiversity of the forest. ✓

> **Tip**
>
> 3-mark questions always have six answer lines – a good indication of how much you should write!

> **Worked example**
>
> Explain **one** way in which volcanoes can form along constructive plate margins.
>
> **A** **(3 marks)**
>
> As the plates pull apart, a plume of magma rises ✓ to the surface to fill the gap. At constructive plate margins this is often basaltic lava which flows easily **B** away from the margin ✓ before solidifying. As more lava rises another layer of rock is formed on top of the first to form a shield volcano. ✓ **C**

> **Examiner feedback**
>
> This candidate makes three separate statements, which are highlighted and marked **A**, **B** and **C**.
>
> This is a good example of a 'chain of reasoning'. Notice how the sentences form a sequence or chain – **A** happens, followed by **B**, followed by **C**. The sequence is:
>
> **A** the rising plume of magma (1 mark)
>
> **B** the basaltic lava flowing away from the margin (1 mark)
>
> **C** the layers building up to form a volcano (1 mark).
>
> So this candidate gets 3 marks.

> **Tip**
>
> When you're explaining how one thing leads to another, you should use **connectives**. These are words linking sentences or phrases together. They include:
>
> - 'because'
> - 'therefore'
> - 'so that'
> - 'which leads to'
> - 'which means that'
>
> See how many you can use in the answers to the questions on the next page!

Now try these!

1   Explain **one** difference between the type of volcanoes found at constructive
    plate boundaries and those found at destructive plate boundaries.

(3 marks)

_____

_____

_____

_____

_____

_____

2   Explain how **one** change in land use can increase the risk of river flooding.

(3 marks)

_____

_____

_____

_____

_____

_____

3   Explain **one** impact of economic development on the rate of urbanisation
    in either a developing or an emerging country.

(3 marks)

_____

_____

_____

_____

_____

_____

On your marks!

4 Managing the statistics questions

4.1 Tackling statistical questions

Ten per cent of all marks in GCSE Geography are for maths and statistical skills. Data are essential to Geography, so you need to be able to handle data and make some fairly straightforward calculations.

This section will give you practice at calculating:

- percentages
- measures of central tendency (a value that represents the centre of a set of data), usually known as mean, median and mode.

Remember, you're allowed to use a calculator in a Geography exam. That will help you, but you do need to show your workings for many questions.

What are the common mistakes?

Candidates often make the following mistakes in exam questions:

- When calculating percentages, candidates often get the data the wrong way round, especially when trying to calculate changes over a period of time.
- They often mix up mean, median and mode.

 Tip

- For 1-mark questions, you don't need to show any working. It's the answer that will get you a mark.
- For 2-mark questions, you get 1 mark for the answer, and the second mark for showing your working. So never just work out the answer on a calculator – you'll lose out on a mark!

Worked example

Study **Figure 1** below. It shows mean monthly temperatures in a city in southern England.

Figure 1

Mean monthly temp. (°C)	Jan	Feb	Mar	Apr	May	Jun	Jul	Aug	Sep	Oct	Nov	Dec
	5	5	7	10	13	15	18	17	15	10	7	5

Calculate the modal value of the mean monthly temperature shown in **Figure 1**.

(1 mark)

Two candidates gave their answers as follows.

Answer 1: _____5_____

This answer is correct. The modal value means the number that occurs most. In **Figure 1**, the most commonly occurring number is 5.

Answer 2: _____10.58_____

This answer is wrong. Look back at **Figure 1** and you might work out where the candidate made a mistake. This answer is the mean temperature, not the modal value. So the candidate clearly confused the two terms.

 Tip

Mean = Add up all the values and then divide by the number of values

Median = The middle value in a ranked data set

Mode = The most common value

Now try these!

1 **Figure 1** shows climate data for central Mali.

Figure 1a (temperature)

	Jan	Feb	Mar	Apr	May	Jun	Jul	Aug	Sep	Oct	Nov	Dec
Temp. (°C)	21.2	24	27.4	30.7	33	33.2	31.8	30.3	30.3	29.4	25.8	21.9

Figure 1b (rainfall)

	Jan	Feb	Mar	Apr	May	Jun	Jul	Aug	Sep	Oct	Nov	Dec
Rainfall (mm)	0.1	0.6	1.6	6.3	16.5	40.8	83.5	114.2	64	16.3	1.3	0.2

Calculate the following:

(a) The mean annual temperature.

(1 mark)

_____

(b) The annual range in temperature.

(1 mark)

_____

(c) The annual total rainfall.

(1 mark)

_____

(d) The percentage of rain that falls during the rainy season of June–September inclusive. Give your answer to one decimal place.

(3 marks)

Show your working.

Percentage of rain = _____ %

2 Study **Figure 2**. It shows changes in economic data for India between 1991 and 2020.

Figure 2

	India 1991	India 2020
GDP total (US$) in PPP	1.2 trillion	8.4 trillion
GDP per capita (US$) in PPP	1150	6100
Exports value (US$)	17.2 billion	485 billion
Imports value (US$)	24.7 billion	493 billion

(a) By how many times did India's GDP increase between 1991 and 2020?

(1 mark)

_____

(b) Calculate the value of the difference between exports and imports in 2020.

(1 mark)

_____

4.2 Manipulating statistics

Exam questions that ask you to manipulate data are common in GCSE Geography. This means processing the data and going beyond making a few calculations such as means or percentages.

These questions are harder because you have to think about the data and know some specialist terms. You could be asked to work with:

- quartiles and inter-quartile ranges
- percentiles
- two sets of data, usually to see whether there's a relationship
- data to make predictions and calculations based on trends (known as extrapolation).

What are the common mistakes?

A surprising number of students get no marks in more complex statistical exam questions. Percentiles, quartiles and inter-quartile ranges are only taught in higher-tier Maths, so you may have studied these for the first time in your GCSE Geography course.

Make sure you understand the terminology at least. Statistical skills are included in the *GCSE (9-1) Geography Edexcel B Student Book*:

- Quartiles and inter-quartile ranges can be found on pages 199, 209, 219 and 229.
- Mean, median and mode are also on pages 199, 209, 219 and 229.

> ### Worked example
>
> 1   A group of students carried out some fieldwork to test the relationship between soil depth and plant height. **Figure 1** shows their results.
>
> *Figure 1*
>
> | Site number | Soil depth (cm) | Plant height (cm) |
> |:---:|:---:|:---:|
> | 1 | 0.0 | 4.0 |
> | 2 | 3.2 | 1.5 |
> | 3 | 3.6 | 6.0 |
> | 4 | 1.9 | 11.5 |
> | 5 | 10.1 | 22.0 |
> | 6 | 15.2 | 65.0 |
> | 7 | 20.2 | 92.0 |
> | 8 | 23.8 | 103.0 |
> | 9 | 32.0 | 129.0 |
> | 10 | 32.0 | 187.4 |
>
> (a) Name **one** type of graph that the students could use to see whether there is a relationship between soil depth and plant height.
>
> (1 mark)
>
> A scattergraph ✓
>
> (b) Explain how you would draw this graph.
>
> (3 marks)
>
> I would draw two axes ✓, one for soil depth and the other for plant height ✓, so that I could then plot the points for each of the ten places. ✓

Examiner feedback

The candidate's answer to **(a)** is correct because a scattergraph is used to find a relationship between two sets of data. (If the candidate had answered 'bar graph' then this would be incorrect because bar graphs are used to show single variables, e.g. the number of cars that passed by in a traffic count.)

For **(b)** the candidate also gets full marks for explaining all three stages in drawing a scattergraph.

Now try these!

1 Students took 16 samples of river sediment at different points downstream.

Study **Figure 1**, a dispersion graph, which plots their data.

 Tip

Remember that 'quartile' means 'divided into four'. Here, for example, the top 4 samples are called the 'upper quartile' and the lower 4 samples are the 'lower quartile'.

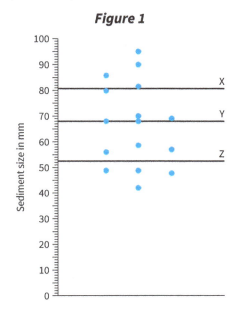

Figure 1

(a) On the graph, three values are labelled X, Y and Z. Complete the table below by writing the correct letter against each term.

(2 marks)

Value	Letter
Median	
Lower quartile	
Upper quartile	

(b) Identify **two** values on **Figure 1** which could be the modal value.

(2 marks)

_____ and _____

5.1 4-mark questions using a resource

In this section you'll learn how to maximise marks on 4-mark questions that use a resource, such as a photo or table of data. 4-mark questions:

- involve writing short paragraphs. To succeed with questions worth 4 or more marks, you must write in full sentences.
- are marked using point marking, like questions carrying 1–3 marks. This means you must develop points using the same guidance as on page 20.

What are the common mistakes?

Like the question below, many 4-mark questions begin with the words 'Using **Figure 1** and your own understanding ...' Many candidates use one or the other – but not both! If you only use **either** Figure 1 **or** your own understanding, you can only gain a maximum of 2 marks. To give you the best chance of receiving high marks, make sure you mention features from Figure 1 **as well as** your own knowledge. First, look at the question below.

Questions using 'suggest'

Study **Figure 1**. It shows damage done by Typhoon Haiyan in 2013.

Figure 1

Using **Figure 1**, suggest **two** reasons why tropical storms have such a big impact in developing and emerging countries.

(4 marks)

For this 4-mark 'suggest' question, you will:

1 Plan your answer
2 Mark an answer
3 Mark a different answer

 Tip

Always 'spot the geography' in the question first – it may help to settle you into thinking about the question. In this example it's 'tropical storms'.

Before attempting to answer the question, remember to **BUG** it. That means:

✓ **Box** the command word, as shown on page 35.

✓ **Underline** the following:

- The focus of the question
- The evidence you need to answer the question
- The number of reasons you need to give for 4 marks.

✓ **Glance** back over the question to make sure you included everything in your answer.

Worked example

Use 'BUGs' like this one to plan your own answers.

Evidence: Support your answer with two things you can see (i.e. evidence) in the photo. You can use your understanding of tropical cyclones as well.

Box and explain the command word: 'suggest' means give intelligent reasons based on the evidence.

Using **Figure 1**, suggest **two** reasons why tropical storms have such a big impact in developing and emerging countries (LICs).

(4 marks)

Focus: The question asks for two reasons about developing and emerging countries (not developed countries!).

What you have to write: You must give two reasons why tropical storms in developing and emerging countries have such an impact. These could be, for example, poverty or the quality of houses.

1 Plan your answer

Plan your answer using the format below. 4-mark answers like this need two developed points. This means:

(a) Picking out two pieces of evidence from Figure 1.
(b) Suggesting a likely reason for each piece of evidence.

(a) Evidence
Identify **two** things you can see in the photo that look like they were caused by the tropical cyclone (i.e. impacts).

1 _____

2 _____

(b) Explanation
Give a possible or likely reason for each of the things you can see in the photo.

1 _____

2 _____

The command word 'suggest' means you aren't expected to know about this tropical cyclone or the Philippines, but that you can think of possible reasons why this tropical cyclone might have had the impacts that it did.

 Tip

The most common mistake in 4-mark questions is that many candidates list only one thing in the photo when the question needs two. This means they're limiting themselves to two marks maximum.

2 Mark an answer

Read through the sample answer below and decide whether it's a good answer or not. Do this by following these steps.

(a) Pick out whether it includes any good points, evidence and explanations. Highlight or underline any:

- evidence in one colour
- explanations in another.

Question recap

Using **Figure 1** (page 34), suggest **two** reasons why tropical storms have such a big impact in developing and emerging countries.

The photo shows wooden buildings which have collapsed, probably because of the high winds in a tropical cyclone. The people who live there are probably poor and have no resistance to storms like this.

In the photo, the different building materials (bits of wood, bamboo) show the building was probably cheap to build, but weak. This is because there might not be inspections or building regulations.

(b) Use the mark scheme below (just like examiners do) to decide what mark to give. In 4-mark questions like this one, examiners would look for two developed points – two points using evidence from the photo, and two explanations about the likely causes.

Mark scheme

Look for two developed points which pick out evidence from the photo and suggest reasons for the impact of the tropical cyclone on the area shown.

Evidence might include

- buildings have collapsed (1)
- wooden building materials / corrugated roofs (1)
- weak building materials (1)
- few visible strong building supports (1)

Reasons might include

- vulnerability of people living in the area / poverty (1)
- lack of government regulations about building quality (1)
- strength of winds in a tropical cyclone (1)

Example answer:

- The buildings look weak and have collapsed during the cyclone (1) because there are few building regulations (1).
- The photo shows wooden housing (1), which could not stand up to the high winds in a tropical cyclone (1).

(c) Fill in the marking table below showing the strengths and weaknesses of the example answer above.

Strengths of this answer	
Ways to improve this answer	
The mark I would give this answer is…	

Worked example

The sample answer is marked below. The text has been coloured to show the strengths of the answer, showing:

- evidence in blue
- likely explanations in yellow.

Evidence: buildings which have collapsed

Explanation: caused by the high winds in a tropical cyclone

Explanation: poverty gives people little resilience against storms

The photo shows wooden buildings which have collapsed probably because of the high winds in a tropical cyclone. The people who live there are probably poor and have no resistance to storms like this.

In the photo, the different building materials (wood, corrugated metal) show the building was probably cheap to build, but weak. This is because there might not be inspections or building regulations.

Evidence: different building materials

Explanation: buildings are weak because there are no inspections or regulations

Examiner feedback

This is a good example of a top quality answer.

- The answer explains the link between the high winds in a tropical storm and the quality of the building. This shows that the candidate has used evidence in the photo.
- The answer explains the weather conditions found during a tropical storm, and how this might affect houses built from cheap, weak building materials. This shows the candidate's understanding.
- In both parts of the answer, the candidate refers directly to the photo.

This answer earns all 4 marks.

3 Mark a different answer

Now you've marked an answer with some guidance, try marking the answer below (this is answering the same question that can be found on page 34).

- Use the mark scheme and highlight the answer like you did before, using the same colours.
- The mark that the answer got from the examiner is in the answers section at the back of this book.

> Wooden shelters like the one in the photo would not be able to stand up to strong hurricane winds so they would fall down.
>
> People in poor countries often live in houses like this on land that isn't theirs.

Questions using 'explain'

Now use the stages you followed to answer the question on page 34 to tackle a different 4-mark question.

Figure 1 shows three measures of development for three countries.

Figure 1

Country	HDI	Death rate per 1000 population	Percentage of population with access to safe water
Japan	0.891	9.51	100
Brazil	0.755	6.58	98
Zimbabwe	0.509	10.13	77

Explain **two** ways in which HDI data in **Figure 1** can help to understand a country's level of development.

(4 marks)

For this 4-mark 'explain' question, you will:

1 **Plan your answer**

2 **Write your answer**

3 **Mark your answer**

4 **Mark a different answer**

5 **Improve an answer**

1 Plan your answer

BUG your answer!

Before attempting to answer the question, remember to **BUG** it. Use the guidelines on page 35 to annotate the question in the boxes below.

Evidence:

Box and explain the command word:

Explain **two** ways in which HDI data in **Figure 1** can help to understand a country's level of development.

(4 marks)

Focus:

What you have to write:

1 Plan your answer

Plan your answer using the format below like the one on page 35. You need to:

(a) Use evidence from the table (Figure 1).

(b) Explain how HDI data can help to understand a country's level of development.

(a) Evidence

State **two** pieces of information that HDI tells us.

1 _____

2 _____

(b) Explanation

Explain what these two pieces of information tell us about a country's level of development.

1 _____

2 _____

The command word 'explain' means you are expected to know about HDI and what it measures as a part of your learning.

2 Write your answer

Explain **two** ways in which HDI data in **Figure 1** can help to understand a country's level of development.

(4 marks)

1 _____

   _____

   _____

   _____

2 _____

   _____

   _____

   _____

Tip

For this question, check that you have written about two ways, not just one.

3 Mark your answer

(a) To help you to identify if your answer includes well-structured points, highlight the:

- evidence in blue
- explanations in orange.

(b) Use the mark scheme below (just like examiners do) to decide what mark to give. In 4-mark questions like this one, examiners would look for two developed points – each explaining what HDI is, and what it tells us about a country's level of development.

Mark scheme

HDI is:

- a single figure (or index) which helps to show a country's spending on welfare / its population (1)

It uses data on the following:

- health / life expectancy (1)
- education / literacy rates (1)
- length of schooling (1)
- GDP / Gross Domestic Product (1)

Explanations about what HDI tells us about a country's level of development include:

- health / life expectancy is about the level of healthcare / how much is spent on healthcare, doctors etc. / how healthy a population is as a whole (1)
- education / literacy rates / length of schooling shows how much a country spends on schools / how well educated a country is / how well people are able to read and write (1)
- GDP / Gross Domestic Product shows how much a country earns on goods and services (1)

Example answer:

- HDI tells us about literacy rates (1) which help to show whether a country can afford to spend money on education (1).
- HDI includes GDP (1), which tells us how much a country earns (1).

(c) Fill in the marking table below, showing the strengths and weaknesses of your answer.

Strengths of my answer	
Ways to improve my answer	
The mark I would give my answer is...	

 Tip

To reach the top marks, you must:

a) show that you know the meaning of HDI and how it is compiled

b) be able to explain how each component of HDI tells us about a country's level of development.

4 Mark a different answer

Read through the sample answer below. This answers the same question.

(a) Annotate the answer with the two colours used above.

HDI is a good measure of a country's development, because it shows how well developed a country is socially as well as economically. It is a single figure that combines GDP (to show how wealthy a country is) with literacy (which shows the level of education) and infant mortality (which shows the level of healthcare). So it is a good way of showing how much money is spent on health and education.

(b) Use the mark scheme on page 40 to decide how many marks the answer is worth.

(c) Fill in the marking table below, showing the strengths and weaknesses of the sample answer above.

Strengths of the answer	
Ways to improve the answer	
The mark I would give the answer is…	

Question recap

Explain **two** ways in which HDI data in **Figure 1** (page 38) can help to understand a country's level of development.

5 Improve an answer

One candidate wrote this answer to the same question. It earns 1 mark. Continue writing so that the answer earns 4 marks.

Life expectancy can measure some things about a country's development, because if people's health is poor then they will die.

_____
_____
_____
_____
_____
_____

Now try this!

Figure 1 shows a map of the geology and rock resistance of a part of the Dorset coast in southern England.

Figure 1 Geology and coastal landforms of southern Dorset

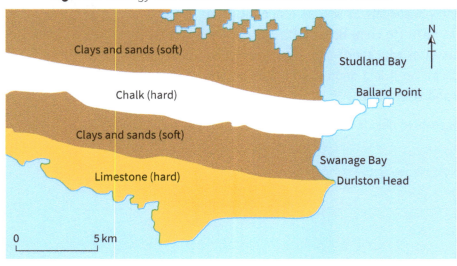

Suggest two ways in which geology has influenced coastal landforms along the coast shown in **Figure 1**. Use **Figure 1** and your own understanding.

(4 marks)

1 _____

_____

_____

_____

2 _____

_____

_____

_____

> **Tip**
>
> Follow each stage in sections 4 and 5 on page 41 to help you tackle this 4-mark question.

> **Tip**
>
> To answer this question fully, check that you know the meanings of these words or phrases:
> - 'geology'
> - 'influenced'
> - 'coastal landforms'.
>
> Also, check where you can see coastal landforms (such as headlands and bays) on the map.

5.2 4-mark questions testing your understanding

In this section you'll learn how to maximise marks on 4-mark questions that do not use resources directly. They depend much more upon your own knowledge and understanding for a successful answer.

Like other 4-mark questions, they:

- involve two short paragraphs of writing , each for 2 marks
- involve two developed explanations, as the question below shows.

Examiners set this kind of question when they want you to develop a point in more detail to show your understanding.

- Think of this just as if you were writing answers to two 2-mark questions.
- This kind of development is exactly the same as in 2-mark and 3-mark questions.

Worked example

Explain two features formed along divergent plate boundaries.

(4 marks)

To answer this question, you would need to name two features found along a divergent plate boundary, and explain each one, as in these examples:

1 There are shield volcanoes because the lava is very runny.

2 Lava flows from within cracks because the crust is being pulled apart.

The answer is made up of two developed points – each naming a feature with an explanation.

Get to know the mark scheme

For the question above, you have to demonstrate:

- knowledge of features formed along a divergent margin
- an understanding of the reasons why these features form.

In the mark scheme below, notice how both knowledge and understanding are needed – for example, a list of four features without any explanation would earn just 2 marks. A weaker answer would show some knowledge of the term 'divergent margin', but would not really show understanding of why features form there. It might even confuse a 'divergent' and 'convergent' margin.

Mark scheme

Look for two features formed along a divergent margin and reasons for their formation.

Example answers:

- A divergent margin forms a crack in the Earth's crust (1), because of the forces pulling the plates apart (1).
- Shield volcanoes form (1) because the lava is hot and runny and flows a long way before cooling (1).

Questions using 'explain'

> Explain how **two** hard methods of coastal engineering can protect the coastline.
>
> **(4 marks)**

For these two 4-mark 'explain' questions, you will:

1 Plan your answer

2 Mark an answer

3 Answer a different question

4 Mark your answer

 1 Plan your answer

(a) Using the boxes below, write out two methods and an explanation for each one about how it protects the coastline.

Method 1:	Method 2:
How it protects the coastline:	How it protects the coastline:

(b) Make sure you can name two methods of hard engineering. Don't confuse them with soft methods!

(c) It's important that you explain how each method protects the coast. Don't just describe more about the method and what it looks like.

Example answers:

- Wooden / stone groynes protect the coast (1) because sand builds up behind them and wave energy is absorbed before a wave reaches a cliff (1).

- A sea wall helps to protect a coastal resort (1) because it reflects the wave energy back out to sea, therefore stopping erosion (1).

 2 Mark an answer

Mark this answer using your mark scheme above. Annotate the answer with your own 'Examiner's feedback'.

1) Beach groynes trap sand brought by longshore drift, which helps to increase beach size and stop erosion.

2) Sea walls also protect the coast but don't absorb energy and so don't stop erosion.

3 Answer a different question

Use the guidance in section 1 on page 44 to help you write your own answer to a different 4-mark question.

Explain two ways in which urban land use can lead to a high risk of flooding.

(4 marks)

1 _____

_____

_____

_____

_____

2 _____

_____

_____

_____

_____

4 Mark your answer

(a) Use this a mark scheme to show what an examiner would look for in an answer to this question:

Your answer should name two ways urban land use can lead to a high risk of flooding, and then explain them.

Example answers:

- Urban land uses are usually impermeable (1) which means water cannot soak into the ground (1).
- Building materials like tarmac or concrete (1) mean water runs off the surface quickly (1).
- There are more drains in urban areas (1) which mean that water can get to rivers very quickly (1).

(b) Mark your answer in section 3 using this mark scheme.

Strengths of my answer	
Ways to improve my answer	
The mark I would give my answer is…	

5.3 Extra practice questions

Now try these!

1 Study **Figure 1**, which shows the location of derelict land and the most deprived areas of Glasgow.

Tip

Use the guidance in section 1 on page 44 to help you answer these questions.

Figure 1

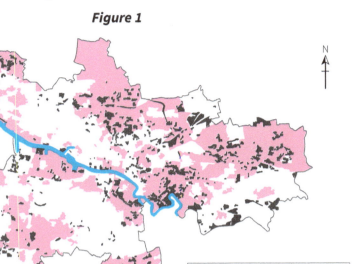

Key
Derelict land
Glasgow's most deprived areas

0   2.5   5 km

Using **Figure 1**, suggest **two** reasons why derelict land and the most deprived areas of Glasgow can be found in similar parts of the city.

(4 marks)

1 _____

_____

_____

_____

2 _____

_____

_____

_____

2 Study **Figure 2**, showing low-income housing in a megacity in one of the world's emerging countries.

Figure 2

Using **Figure 2**, explain **two** ways in which urban growth in emerging countries has affected quality of life.

(4 marks)

1 _____

_____

_____

_____

2 _____

_____

_____

_____

3 Study **Figure 3**, which shows ice coverage in the Arctic.

Figure 3 Arctic sea ice sets record low

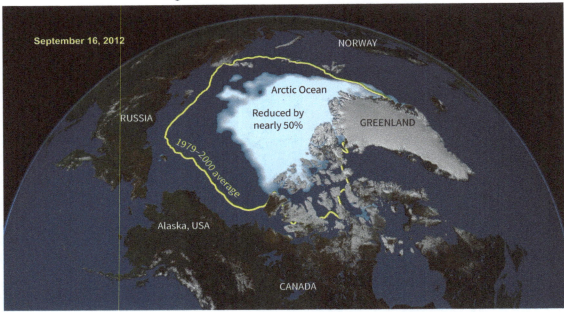

Explain **two** pieces of evidence showing that climate change has affected the region in **Figure 3**.

(4 marks1

1 _____

_____

_____

_____

2 _____

_____

_____

_____

 Tip

When using evidence to explain something in your argument, refer directly to the figure that you are given. For example, in question 3, you should refer to place names and where the ice cap seems to have retreated, or where it has stayed the same.

 **Review**

Review your answers to 'Now try these!' questions 1, 2 and 3 on pages 46–48 by completing the marking tables below.

Question 1

Strengths of my answer	
Ways to improve my answer	
The mark I would give my answer is…	

Question 2

Strengths of my answer	
Ways to improve my answer	
The mark I would give my answer is…	

Question 3

Strengths of my answer	
Ways to improve my answer	
The mark I would give my answer is…	

- **In this section you'll learn how to tackle 8-mark questions using command words 'Assess' and 'Evaluate'.**

8-mark questions differ from 4-mark questions:

- They have tougher command words such as 'Assess' or 'Evaluate'. 4-mark questions tend to use 'Explain' or 'Suggest'.
- They are marked using levels, not points. Levels are criteria (qualities) which examiners look for and use to judge your answer. Level 1 is the lowest standard (1–3 marks) and Level 3 is the highest (worth 7–8 marks).

What does 'assess' mean?

'Assess' means using evidence to decide how significant something is. In a list of possible features or explanations, you need to decide which are the most important. For example:

Assess the effectiveness of hard engineering as a way of managing coastlines.

To 'assess' the effectiveness of hard engineering, you would have to:

- list three methods of hard engineering that you know (three is enough for an 8-mark answer)
- weigh up how effective each one is
- decide which are most effective and why.

What does 'evaluate' mean?

'Evaluate' means weighing up e.g. an opinion, or set of factors, and providing a judgment or conclusion on the basis of evidence, e.g. strengths and weaknesses, or relevant data. For example:

Evaluate the effects of economic development on one named city in either a developing or emerging country.

To 'evaluate' the effects of industrialisation on a city, you have to:

- show that you know about three different effects of industrialisation on a city
- show evidence or examples of these three effects on a named city
- explain how some effects may be greater than others
- make a judgment – which are the greatest / less serious effects? This would need a short conclusion of one or two sentences.

Examples of 4- and 8-mark questions

A 4-mark question might ask:

Explain two economic impacts of either an earthquake or a volcanic eruption.

An 8-mark question might ask:

Evaluate the statement that 'tectonic events have greater economic impacts on developing and emerging countries than they do on developed countries'.

'Assess' and 'Evaluate' questions in Paper 1

'Assess' and 'Evaluate' questions in Paper 1 are different from those in Paper 2. Those in Paper 1 will not include any stimulus material. They assess:

- your understanding of what you've learned (AO2), which carries 4 marks
- your ability to weigh up, assess, and argue a case (AO3), which carries 4 marks.

Only writing about what you know (AO2) without any assessment or judgment limits you to 4 marks. The mark scheme is shown in Figure 1.

'Assess' and 'Evaluate' questions in Papers 2 and 3

8-mark questions in Papers 2 and 3 use a resource (e.g. photo, table of data) linked to a question, e.g. fieldwork data in Section C1 or C2. Questions ask you about the data, and to develop an argument about something linked to them. The mark scheme is also shown in Figure 1.

This means 8 marks are divided into:

- 4 marks AO3, because you'll make a judgment or assessment linked to the question
- 4 marks AO4, because you'll be using geographical skills in interpreting the resource.

Level	Marks	Descriptor for Paper 1	Descriptor for Papers 2 and 3
	0	No acceptable response	No acceptable response
1	1–3	**For AO2** • Shows isolated elements of understanding of concepts and links between places, environments and processes. **For AO3** • Attempts to apply understanding to deconstruct information but understanding and connections are flawed. • An imbalanced or incomplete argument. • Judgments supported by limited evidence.	**For AO3** • Attempts to apply understanding to deconstruct information but understanding and connections are flawed. • An imbalanced or incomplete argument. • Judgments supported by limited evidence. **For AO4** • Uses some geographical skills to obtain information with limited relevance and accuracy, which supports few aspects of the argument.
2	4–6	**For AO2** • Shows elements of understanding of concepts and links between places, environments and processes. **For AO3** • Applies understanding to deconstruct information and give some logical connections between concepts. • An imbalanced argument that draws together some points. • Judgments supported by some evidence.	**For AO3** • Applies understanding to deconstruct information and give some logical connections between concepts. • An imbalanced argument that draws together some points. • Judgments supported by some evidence. **For AO4** • Uses geographical skills to obtain accurate information that supports some aspects of the argument.
3	7–8	**For AO2** • Shows accurate understanding of concepts and links between places, environments and processes. **For AO3** • Applies understanding to deconstruct information and make logical connections between concepts. • A balanced, well-developed argument that draws together relevant points coherently. • Makes judgments supported by evidence.	**For AO3** • Applies understanding to deconstruct information and make logical connections between concepts. • A balanced, well-developed argument that draws together relevant points coherently. • Makes judgments supported by evidence. **For AO4** • Uses geographical skills to obtain accurate information that supports all aspects of the argument.

Figure 1 *Marking criteria for all 8-mark questions in Paper 1, and Papers 2 and 3, using the command words 'Assess' and 'Evaluate'.*

- **In this section you'll learn how to tackle 8-mark questions which use 'Assess' as a command word in Paper 1.**

Study **Figure 1**. It shows a street scene in a low-income housing area in a developing / emerging country.

Figure 1

'For those who live in low-income areas of megacities in developing or emerging countries, life presents far more problems than benefits.'

Question
Assess the extent to which you agree with this statement.

(8 marks)

Six steps to success!

The following six activity steps are used in this chapter to help you get the best marks.

1 **Plan your answer** – decide how to structure your answer.

2 **P-E-E-L your answer** – plan what you want to say and the points you want to make.

3 **Write your answer** – use the answer spaces to write your answer.

4 **Mark your answer** – use the mark scheme to self- or peer-mark your answer. You can also use this to mark sample answers in step 5 below.

5 **Mark different answers** – sample answers are given to show you how to mark an answer.

6 **Review the marks for different answers** – these are the same answers as for step 5, but are marked and annotated, so that you can compare these with your own.

 1 Plan your answer

Before attempting to answer the question, remember to **BUG** it. On a separate piece of paper, annotate it using the guidelines on pages 34–35.

Assess the extent to which you agree with this statement. (8 marks)

Remember!
- 'Assess' in Paper 1 means using your own knowledge and understanding.
- It assesses AO2 (your understanding) and AO3 (your ability to apply what you understand to the question).
- You can use the photo to help you think about the megacity you have studied.

2 P-E-E-L your answer

Use P-E-E-L notes to help structure your answer. P-E-E-L stands for Point, Evidence, Explanation and Link. It has four stages:

- **P**oint – You need to make **three** points for an 8-mark answer, based on your understanding of megacities e.g. *'Point 1 – I agree, services such as sanitation are often very poor or non-existent.'*
- **E**vidence – This is the evidence you want to use from your understanding of megacities to support Point 1, e.g. *'The photo shows open sewers which contain waste and probably sewage.'*
- **E**xplanation – This is your explanation for the evidence, e.g. *'This is because housing like this is often informal, without services such as sanitation, because the city authorities don't receive any funding to install it.'* Start your explanation with 'This is because …'
- **L**ink – Refer back to the question about whether living in low-income areas of megacities presents more problems than benefits. Finish the final paragraph with a one- or two-sentence conclusion about your judgment.

Tip

Make a judgment!
Don't just describe and explain. If the question asks you to 'Assess', it wants you to make a judgment. Is it true that *life presents far more problems than benefits … for those who live in low-income areas of megacities?* One way of doing this is in a **mini-conclusion** where you make a judgment – it need only be a sentence or two.

Planning grid

- Use this planning grid to help you write high-quality paragraphs. Starter phrases have been used to help you. Remember to include links to show how your points relate to the question 'assess the extent to which you agree'. Remember – **make a judgment**!
- The example below is based on an argument that there are more benefits than problems. But you can alter this if you wish – it's your judgment!

	Paragraph 1	**Paragraph 2**	**Paragraph 3**
Point	The first problem is …	The second problem is …	However, there are also benefits because …
Evidence	The evidence for this is …	The evidence for this is …	The evidence for this is …
Explanation	This is because …	This is because …	This is because …
Link back to the question – how far do you agree?	This shows how life presents more problems than benefits because …	This shows how life presents more problems than benefits because …	So, overall, the problems outweigh benefits (*or vice-versa*) because …

3 Write your answer

'For those who live in low-income areas of megacities in developing or emerging countries, life presents far more problems than benefits.'

Assess the extent to which you agree with this statement.

(8 marks)

_____

_____

_____

_____

_____

_____

_____

_____

_____

_____

_____

_____

_____

_____

_____

_____

Strengths of the answer	
Ways to improve the answer	

Level		Mark	

4 Mark your answer

1   To help you to identify if the answer includes well-structured points, first highlight or underline the:

- points in red   • explanations in orange   • evidence in blue
- points which show the candidate is assessing the statement. These might support one side of the argument, or balance it before reaching a conclusion.

2   Use the mark scheme below. 8-mark questions are not marked using individual points. Instead, choose a level and a mark based upon the quality of the answer as a whole.

Level	Marks	Descriptor	Examples
	0	No acceptable response	
1	1–3	**For AO2** • Shows isolated elements of understanding of concepts and links between places, environments and processes. **For AO3** • Attempts to apply understanding to deconstruct information but understanding and connections are flawed. • An imbalanced or incomplete argument. • Judgments supported by limited evidence.	• *Developing cities have no water or sewage pipes and have many health problems from drinking bad water.* • *There are so many people that the city cannot keep pace with them all.* • *So the statement is right because life there is very hard and the city cannot support all those people.*
2	4–6	**For AO2** • Shows elements of understanding of concepts and links between places, environments and processes. **For AO3** • Applies understanding to deconstruct information and give some logical connections between concepts. • An imbalanced argument that draws together some points. • Judgments supported by some evidence.	• *Cities like Mumbai often have few water or sewage connections, and electricity in the photo looks unsafe too.* • *This is because many people are poor and cannot afford water or electricity bills.* • *So the statement is true because most people do not have a decent lifestyle with basics that people in developed countries take for granted.*
3	7–8	**For AO2** • Shows accurate understanding of concepts and links between places, environments and processes. **For AO3** • Applies understanding to deconstruct information and make logical connections between concepts. • A balanced, well-developed argument that draws together relevant points coherently. • Makes judgments supported by evidence.	• *In Mumbai, a third of homes have no electricity (or have illegal hook-ups from wires) and half have no sewage connections.* • *One reason is that suburbs are growing so quickly that the city council cannot keep pace with population growth.* • *This shows the statement is true because electricity and sewage connection are basics for a reasonable life. But there are benefits, such as provision of schooling.*

Read through Sample answers 1 and 2.

a) Go through each one using the three colours in section 4. Remember to underline points that show the candidate is assessing.

b) Use the level descriptions to decide how many marks each one is worth.

Sample answer 1

I agree with the statement. Mumbai has doubled in size since 1990. Suburbs like Dharavi are growing so quickly that it is hard to keep pace with services like water. It is better than it was because now houses are being built out of brick instead of timber and odd bits of metal, and many also have water and electricity. There are shops there and many services like health facilities that you would expect. But I agree with the statement because Dharavi is probably one of Mumbai's best suburbs if you're poor and there are many worse that do not have half the benefits that it has. You wouldn't choose to live there if you had more money so areas like that are still for low-income people, so I still think the statement is true.

Elsewhere Mumbai has informal settlements, which are places where people build their own shelters illegally. Some of these shelters are on sloping land because nobody else wants to live there and they can be a long way from jobs in the city centre. But when it rains heavily, people are vulnerable, because in 2021 over 30 people were killed in a landslide. This shows that the statement is true, because the poor have to live there – people with jobs and decent incomes would never choose to live in places like that.

Use a copy of the marking grid below.

Sample answer 2

I don't agree with the statement. It is true that people living in informal settlements have a lot of problems like they don't have water supply or sewerage connections and when you walk down the street then you might be electrocuted as there are often bare wires from illegal hook-ups. But cities have many jobs for people and so the people who have moved there from the countryside are often employed more than if they had stayed in rural areas. Many rural areas do not have schools and cities like Mumbai have plenty of schools for all ages, maybe universities too. There are often hospitals and medical treatment in cities that you don't have in the countryside. So it's not perfect living in Mumbai but it can be better than a lot of places so I don't agree with the statement.

Strengths of the answer	
Ways to improve the answer	
Level		Mark	

6 Review the marks for different answers

Sample answers 1 and 2 are marked below. The following have been highlighted to show how well each answer has structured points.

- **points** in red
- **explanations** in orange
- **evidence** in blue
- <u>judgments</u> are underlined. These are important in order to reach Level 3 on questions whose command word is 'Assess'.

Marked sample answer 1

Point – the candidate quantifies the growth of Mumbai

Explanation – an explanation is given for the impact of this growth

Evidence – the candidate uses the evidence of building materials

Evidence – the candidate uses further evidence of shops and health services

Judgment – the candidate makes a comparison to justify their choice

Judgment – the candidate makes a further statement to justify their choice

I agree with the statement. Mumbai has doubled in size since 1990. Suburbs like Dharavi are growing so quickly that it is hard to keep pace with services like water. It is better than it was because now houses are being built out of brick instead of timber and odd bits of metal, and many also have water and electricity. There are shops there and many services like health facilities that you would expect. But I agree with the statement because Dharavi is probably one of Mumbai's best suburbs if you're poor and there are many worse that do not have half the benefits that it has. You wouldn't choose to live there if you had more money so areas like that are still for low-income people, so I still think the statement is true.

Elsewhere Mumbai has informal settlements, which are places where people build their own shelters illegally. Some of these shelters are on sloping land because nobody else wants to live there and they can be a long way from jobs in the city centre. But when it rains heavily, people are vulnerable because in 2021 over 30 people were killed in a landslide. This shows how the statement is true, because the poor have to live there – people with jobs and decent incomes would never choose to live in places like that.

Explanation – informal settlements are explained

Evidence – the candidate uses evidence about land used by informal settlements

Point – the candidate describes the vulnerability of people

Point – the candidate mentions informal settlements

Judgment – the candidate gives one further supporting statement to justify their choice, though it is very similar to the second judgment

Explanation – an explanation is given to illustrate this

 Examiner feedback

This candidate knows a lot and has a clear view of what living in a city like this might be like. Points are well made and extended points offer detail about informal housing to support the answer. The candidate also clearly justifies why they have reached an opinion. The answer is generally Level 3 in quality. However, it is not perfect, as the answer is short on data. It is Level 3 with sound justification, so it earns 7 marks.

Marked sample answer 2

Point – the candidate makes an illustrated point about informal settlements

I don't agree with the statement. It is true that people living in informal settlements have a lot of problems like they don't have water supply or sewerage connections and when you walk down the street then you might be electrocuted as there are often bare wires from illegal hook-ups. But cities have many jobs for people and so the people who have moved there from the countryside are often employed more than if they had stayed in rural areas. Many rural areas do not have schools and cities like Mumbai have plenty of schools for all ages, maybe universities too. There are often hospitals and medical treatment in cities that you don't have in the countryside. So it's not perfect living in Mumbai but it can be better than a lot of places so I don't agree with the statement.

Evidence – the candidate extends the point using evidence about electricity

Evidence – the candidate uses evidence of employment to compare cities and rural areas

Point – the candidate makes the point about education in cities

Evidence – the candidate uses the evidence of health care to extend the point further

Point – the candidate makes the point about employment in cities

Judgment – the candidate makes a single statement about living in Mumbai, though this is not a quality comparison.

Examiner feedback

This is a medium quality answer which was given 5 marks in the middle of Level 2. The candidate makes three valid points about living in informal settlements and extends it with some detail. Generic writing, without naming a place, is usually typical of Level 1 – so the candidate has saved themselves by naming Mumbai twice. The level of judgment is weak; there is no other named place to compare cities with, simply mentioning rural areas.

This candidate probably knows more than this, so some revision of a named city would have earned higher marks – perhaps with some named examples of a megacity they know, or some data illustrating households with water supply etc. Judgment needs to be more than just a general statement at the end.

On your marks!

6.3 8-mark questions using 'Assess' in Papers 2 and 3

- In this section you'll learn how to tackle 8-mark questions which use 'Assess' as a command word in Papers 2 and 3. These use a resource which you must refer to, using your geographical skills (AO4).

Study **Figure 1**, which is a map showing the distribution of Asian Indian British people in London.

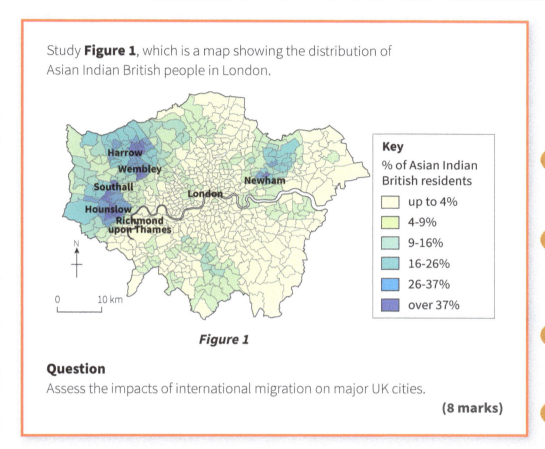

Key

% of Asian Indian British residents

- up to 4%
- 4-9%
- 9-16%
- 16-26%
- 26-37%
- over 37%

Figure 1

Question

Assess the impacts of international migration on major UK cities.

(8 marks)

Six steps to success!

The following six activity steps are used in this chapter to help you get the best marks.

1 **Plan your answer** – decide how to structure your answer.

2 **P-E-E-L your answer** – plan what you want to say and the points you want to make.

3 **Write your answer** – use the answer spaces to write your answer.

4 **Mark your answer** – use the mark scheme to self- or peer-mark your answer. You can also use this to mark sample answers in step 5 below.

5 **Mark different answers** – sample answers are given to show you how to mark an answer.

6 **Review the marks for different answers** – these are the same answers as for step 5, but are marked and annotated, so that you can compare these with your own.

1 Plan your answer

Before attempting to answer the question, remember to **BUG** it. On a separate piece of paper, annotate it using the guidelines on pages 34–35.

Question

Assess the impacts of international migration on major UK cities.

(8 marks)

Remember!

- 'Assess' is assessed in Papers 2 and 3 using a resource stimulus, like the map in this example.
- It assesses AO4 (your skill in interpreting the map) and AO3 (your ability to apply what you see to the question).
- You can use what you know about cities as well as what's in the resource.

2 P-E-E-L your answer

Use PEEL notes to structure your answer. This will help you to communicate your ideas to the examiner in the clearest way. PEEL has four stages:

- **P**oint – You need to make **three** points for an 8-mark answer, based on the map in Figure 1, e.g. *'Point 1 – Asian Indian British residents are concentrated in certain areas of cities.'*
- **E**vidence – Include specific evidence (e.g. names of suburbs) from the map and other major UK cities that you know about.
- **E**xplanation – This is your explanation for the evidence, e.g. *'This is because migrant communities live close by for reasons of security or religion.'* Start your explanation with *'This is because ...'*
- **L**ink – Link back to the question about the impacts of international migration on major UK cities. Finish with a one- or two-sentence conclusion about how big the impacts have been.

Tip

Make a judgment! Don't just describe and explain. If the question asks you to 'Assess', it wants you to make a judgment. How great have the *impacts of international migration* been *on cities in the UK*? One way of doing this is in a **mini-conclusion** where you make a judgment – it need only be a sentence or two.

Planning grid

Use this planning grid to help you write high-quality paragraphs. Use the guidance in section 6.2 (page 53) to help you.

	Paragraph 1	Paragraph 2	Paragraph 3
Point			
Evidence (*from map or your own knowledge*)			
Explanation			
Link – a mini conclusion			

3 Write your answer

Assess the impacts of international migration on major UK cities.   **(8 marks)**

_____

_____

_____

_____

_____

_____

_____

_____

_____

_____

_____

_____

_____

_____

_____

Strengths of the answer			
Ways to improve the answer			
Level		**Mark**	

🖉 4 Mark your answer

1 To help you to identify if the answer includes well-structured points, first highlight or underline:

- points in red • explanations in orange • evidence in blue
- any judgments.

2 Use the mark scheme below to decide what mark to give. 8-mark questions are not marked using individual points, but instead you should choose a level and a mark based upon the quality of the answer as a whole.

Level	Marks	Descriptor	Examples
	0	No acceptable response	
1	1–3	**For AO3** • Tries to apply understanding but understanding and connections are flawed. • Imbalanced or incomplete argument. • Judgments supported by limited evidence. **For AO4** • Uses some geographical skills to obtain information with limited relevance and accuracy.	• *London has a lot of immigrants living there so the city is growing.* • *Immigrants often live in the same sorts of areas where they have familiar shops or mosques, and they like living there.*
2	4–6	**For AO3** • Applies understanding to deconstruct information and give some logical connections between concepts. • An imbalanced argument that draws together some points. • Judgments supported by some evidence. **For AO4** • Uses geographical skills to obtain accurate information that supports some of the argument.	• *Immigration has been the reason for half the recent growth of cities such as London. In London there are over 200 languages spoken in the city now.* • *Figure 1 shows that immigrants often settle in suburbs where there are cultural or ethnic groups like their own.*
3	7–8	**For AO3** • Applies understanding to deconstruct information and make logical connections between concepts. • A balanced, well-developed argument that draws together relevant points coherently. • Makes judgments supported by evidence. **For AO4** • Uses geographical skills to obtain accurate information that supports most of the argument.	• *The impacts of immigration have been great especially on the culture of UK cities. In areas like Southall on the map, tourists might be attracted to curry restaurants or shopping at a range of Asian food shops.* • *Figure 1 shows that immigrants from particular countries, religions or cultures tend to live in areas close to each other, creating suburbs like Southall in west London.*

5 Mark different answers

Read through Sample answers 1 and 2.

a) Go through each one using the three colours in section 4. Remember to underline any **judgments**, because these are needed to meet the requirements of the command word 'Assess'.

b) Use the level descriptions to decide how many marks it is worth.

Sample answer 1

Half of London's population growth in recent years has been because of migrants from overseas, from countries such as India in Figure 1, because of the jobs available there such as in construction and financial services. International migration has affected London because people from over 200 countries have settled there. Over 37% of the people living in Newham are Asian Indian British residents.

Like the map in Figure 1, many immigrants have changed the character of the parts of the city where they live, because you would probably find shops or places of worship in areas like Wembley. This changes the culture in cities and there are festivals like the Notting Hill Carnival in London. So immigration has had a big effect on cities.

Strengths of the answer	
Ways to improve the answer	
Level		Mark	

Sample answer 2

Cities like London are growing fast because of immigrants from other countries. There are jobs in London which attract people to live there. It has meant that there is pressure on jobs and housing but London gains because there are also new restaurants and festivals which helps the city's image. When migrants arrive they look for work anywhere, but when they get jobs, their families come and join them, so that's what makes the population go up so quickly.

Strengths of the answer	
Ways to improve the answer	
Level		Mark	

6 Review the marks for different answers

Sample answers 1 and 2 are marked below and opposite. The following have been highlighted to show how well each answer has structured points:

- points in red   • explanations in orange   • evidence in blue
- judgments are underlined. These are important in order to reach Level 3 on questions whose command word is 'Assess'.

Marked sample answer 1

Point – candidate quantifies the amount of population growth due to immigration

Evidence – the example of India illustrates the point

Explanation – the reason is given for immigration, i.e. employment (with examples)

Point – candidate quantifies the extent of immigration from different countries

Evidence – evidenced with the example of Newham data from Figure 1

Point – the point helps to answer the part of the question dealing with changing character of cities

Half of London's population growth in recent years has been because of migrants from overseas, from countries such as India in Figure 1, because of the jobs available there such as in construction and financial services. International migration has affected London because people from over 200 countries have settled there. Over 37% of the people living in Newham are Asian Indian British residents.

Like the map in Figure 1, many immigrants have changed the character of the parts of the city where they live, because you would probably find shops or places of worship in areas like Wembley. This changes the culture in cities and there are festivals like the Notting Hill Carnival in London. So immigration has had a big effect on cities.

Explanation – the candidate explains how the character is changed, with examples

Judgment – the candidate makes a judgment about how immigration changes the city. This is not a very strong judgment – but it does fit the command word 'Assess'.

Evidence – candidate illustrates the point with an example

 Examiner feedback

The descriptors for Level 3 apply to this answer as follows:

For AO3:

- *Applies understanding to deconstruct information and make logical connections between concepts.*
- *A balanced, well-developed argument that draws together relevant points coherently.*
- *Makes judgments supported by evidence.*

The candidate has been able to mention both the reasons for growth and the changing character of London. A specific source country is named, data are used, and there is an example of the kind of cultural events resulting from immigration. It does not matter that these examples are chosen from London.

For AO4:

- *Uses geographical skills to obtain accurate information that supports most of the argument.*

This is not so strong. The candidate refers to Figure 1, but only briefly.

By meeting the first descriptor fully, and the second one partly, the answer is low Level 3 in quality. The judgment is also weaker than would be needed for a top Level 3, so the answer is worth 7 marks.

Marked sample answer 2

Point – a general and non-specific point helps to answer the part of the question dealing with growth of cities

Explanation – the candidate briefly explains the reason for growth, but without specific examples

Explanation – the candidate explains how jobs help to explain immigration but does not offer examples

Cities like London are growing fast because of immigrants from other countries. There are jobs in London which attract people to live there. It has meant that there is pressure on jobs and housing but London gains because there are also new restaurants and festivals which help the city's image. When migrants arrive they look for work anywhere, but when they get jobs, their families come and join them, so that's what makes the population go up so quickly.

Point – this point helps to answer the part of the question dealing with changing character of cities

Evidence – candidate evidences the point by showing the benefit of restaurants and festivals

Explanation – the candidate revisits the explanation for growth of population

Note that there are no judgments made in this answer.

 Examiner feedback

The descriptor for Level 1 applies to this answer, as follows:

For AO3:

- *Tries to apply understanding but understanding and connections are flawed.*
- *Imbalanced or incomplete argument.*
- *Judgments supported by limited evidence.*

The candidate quotes patterns from London and clearly understands how important migration, and immigration particularly, are to explain its rapid growth of population. In the latter part of the answer, the candidate also mentions the importance of family members as a reason for further increase. The candidate understands the impact of immigration in terms of food and festivals, but there are no specific examples.

For AO4:

- *Uses some geographical skills to obtain information with limited relevance and accuracy* – the candidate does not refer to Figure 1 at all in the answer – though there are hints of having looked at it.

By meeting the descriptor for Level 1, the answer gains 3 marks.

On your marks!

6.4 8-mark questions using 'Assess' in Paper 2 Fieldwork

- **In this section you'll learn how to tackle 8-mark questions which use 'Assess' as a command word in the fieldwork section of Paper 2. You must use the examples from fieldwork that you have done, using your geographical skills (AO4).**

You can use **either** your physical **or** your human fieldwork investigation to answer this question.

> **Question**
> Using the conclusions from your geographical investigation, assess the accuracy and reliability of your results.　　　　**(8 marks)**

1 Plan your answer

Before attempting to answer the question, remember to **BUG** it. That means:

✔ **Box** the command word.
✔ **Underline** the following:
   - the theme
   - the focus
   - any evidence required
   - the number of examples needed.
✔ **Glance** back over the question – to make sure you include everything in your answer.

Annotate the question in the space below. Use the example on pages 34–35 to help you.

> Using the conclusions from your geographical investigation, assess the accuracy and reliability of your results.　　　　**(8 marks)**

Remember!
- 'Assess' makes use of your fieldwork experience and skills, when it is used in the fieldwork section of Paper 2.
- It assesses AO3 (your ability to apply what you experienced to the question) and AO4 (your skill in fieldwork).

Six steps to success!

The following six activity steps are used in this chapter to help you get the best marks.

1 **Plan your answer** – decide how to structure your answer.

2 **P-E-E-L your answer** – plan what you want to say and the points you want to make.

3 **Write your answer** – use the answer spaces to write your answer.

4 **Mark your answer** – use the mark scheme to self- or peer-mark your answer. You can also use this to mark sample answers in step 5 below.

5 **Mark different answers** – sample answers are given to show you how to mark an answer.

6 **Review the marks for different answers** – these are the same answers as for step 5, but are marked and annotated, so that you can compare these with your own.

2 P-E-E-L your answer

Use PEEL notes to structure your answer. This will help you to communicate your ideas to the examiner in the clearest way. PEEL has four stages:

- **P**oint – You need to make **three** points for an 8-mark answer, based on the conclusions drawn from your fieldwork, e.g. *'Point 1 – Some of our results were accurate, such as ...'*
- **E**vidence – Include details from either your physical fieldwork or your human fieldwork to support each point.
- **E**xplanation – This is your explanation for the evidence, e.g. *'This is because equipment such as a flow meter is more accurate than using orange peel to measure the speed of a river.'* Start your explanation with *'This is because ...'*
- **L**ink – Link back to the question about the accuracy and reliability of your results. Finish with a one- or two-sentence conclusion about how accurate and reliable you were.

Tip

Make a judgment! Don't just describe and explain. The question is asking you to 'Assess', so it wants you to make a judgment. How *accurate* were your fieldwork results? How *reliable* were they? One way of doing this is in a **mini-conclusion** where you make a judgment – it need only be a sentence or two.

What's the difference between accuracy and reliability?

- **Accuracy** means how carefully you collected results to make sure they were as true as they could be.
- **Reliability** means that if you went back to the same place on a different day or time, you'd get the same results as you did when you visited.

Planning grid

Use this planning grid to help you write high-quality paragraphs. Use the guidance in section 6.2 (page 53) to help you.

	Paragraph 1	Paragraph 2	Paragraph 3
Point			
Evidence (*based on your fieldwork experience*)			
Explanation			
Link – a mini conclusion			

3 Write your answer

Using the conclusions from your geographical investigation, assess the accuracy and reliability of your results.

(8 marks)

Note: *You can use either your physical or your human fieldwork investigation to answer this question.*

_____

_____

_____

_____

_____

_____

_____

_____

_____

_____

_____

_____

_____

_____

_____

Strengths of the answer	
Ways to improve the answer	
Level		Mark	

4 Mark your answer

1 To help you to identify if the answer includes well-structured points, first highlight or underline:

- points in red
- explanations in orange
- evidence in blue
- any judgments about accuracy and reliability.

2 Use the mark scheme below to decide what mark to give. 8-mark questions are not marked using individual points, but instead you should choose a level and a mark based upon the quality of the answer as a whole. You'll have to adapt this when marking Sample answer 2 about urban fieldwork.

Level	Marks	Descriptor	Examples
	0	No acceptable response	
1	1–3	**For AO3** • Tries to apply understanding but understanding and connections are flawed. • Imbalanced or incomplete argument. • Judgments supported by limited evidence. **For AO4** • Few aspects of the enquiry process are supported by the use of geographical skills. • Communicates general fieldwork findings with limited relevance and accuracy and little relevant geographical terminology.	• *We measured width and depth of the river using a tape measure and a ruler.* • *Some of our results weren't accurate because we lost our tape measure and had to use paces.* • *If we went back on another day we'd probably get different results.*
2	4–6	**For AO3** • Applies understanding to deconstruct information and give some logical connections between concepts. • An imbalanced argument that draws together some points. • Judgments supported by some evidence. **For AO4** • Some aspects of the enquiry process are supported by the use of geographical skills. • Communicates fieldwork findings fairly clearly using occasional relevant geographical terminology.	• *We measured the area of the river, then velocity to find volume.* • *Sometimes velocity readings weren't very accurate because our dog biscuit got wet. So I would not trust our volume results as other groups got different results from ours.* • *If we went back on another day it might be drier, so we'd get different results.*
3	7–8	**For AO3** • Applies understanding to deconstruct information and make logical connections between concepts. • A balanced, well-developed argument that draws together relevant points coherently. • Makes judgments supported by evidence. **For AO4** • All aspects of the enquiry process are supported by the use of geographical skills. • Communicates specific fieldwork findings clearly, and consistently uses relevant geographical terminology.	• *Our fieldwork included measuring width and depth of the river, then velocity. Multiplied together, these give discharge calculations, which would vary between one day and another.* • *Our results relied upon each member of each group doing width and depth readings accurately. The group I was with was sometimes a bit rushed, which affected our accuracy.* • *Our readings could not be reliable because the river might have a different discharge on a different day, depending on rainfall.*

5 Mark different answers

Read through Sample answers 1 and 2.
a) Go through each one using the three colours in section 4. Remember to underline any **judgments** about accuracy and reliability, because these are needed to answer the question.
b) Use the level descriptions to decide how many marks it is worth.

Sample answer 1 – rivers fieldwork

We measured width, depth and velocity (to calculate discharge) and gradient. The river floods in places, so we used the wider banks of the river to calculate channel capacity at bank-full so that we could calculate how easily it would flood by looking at the maps of flood risk from the Environment Agency. Our BGS phone app also told us the geology and whether rocks are porous or permeable. Both of these were accurate.

Some widths were hard to measure because banks were slippery, making results less accurate. Depth readings varied across the river, so we took several readings and averaged them. We all measured the same sites, but some group results varied perhaps because they didn't do it carefully enough. We took photos so that we could work out which group's results were the accurate ones.

Equipment like flowmeters didn't always work, and gradient was hard to work out using apps on our phones because we didn't want them to get wet. When we did our fieldwork it had rained the day before, so our results would not be reliable compared to a dry summer day. So, in conclusion, we were only partly accurate, and reliability would be difficult depending on the weather.

Use a copy of the marking grid opposite to assess this answer.

Sample answer 2 – urban fieldwork

Generally our results were correct but only in certain ways. We used an environmental quality survey to find out what people living there thought about their area. As long as everyone recorded the data correctly, it would be accurate. But to get a proper picture of the area, we used the Living Environment IMD data, as that's government data and would be reliable.

We also measured noise levels using decibel apps on our phones and counted traffic, which gave a clearer picture of environmental quality and made our results more reliable. It helped when we plotted noise data on a map, as we could see that the noisiest parts were along the main road. You can also tell our results were accurate because everyone said the same thing about litter and graffiti and that it was poor (which it was). We did get different results from different age groups because we collected data in the morning when people were at school and work, so it was mainly old people in one area. That might make our results wrong or misleading.

Strengths of the answer			
Ways to improve the answer			
Level		**Mark**	

✏ 6 Review the marks for different answers

Sample answers 1 and 2 are marked below. The following have been highlighted to show how well each answer has structured points:

- points in red
- explanations in orange
- evidence in blue
- judgments are underlined. These are important in order to reach Level 3 on questions whose command word is 'Assess'.

Marked sample answer 1

Point – identifies a method of data collection

Explanation – extends the point to compare primary with secondary data

Evidence – data collection by the candidate

We measured width, depth and velocity (to calculate discharge) and gradient. The river floods in places, so we used the wider banks of the river to calculate channel capacity at bank-full so that we could calculate how easily it would flood by looking at the maps of flood risk from the Environment Agency. Our BGS phone app also told us the geology and whether rocks are porous or permeable. Both of these were accurate.

Judgment – links back to the question, though the candidate does not say why the data are accurate

Some widths were hard to measure because banks were slippery, making results less accurate. Depth readings varied across the river, so we took several readings and averaged them. We all measured the same sites, but some group results varied perhaps because they didn't do it carefully enough. We took photos so that we could work out which group's results were the accurate ones.

Point – identifies a problem with data accuracy

Explanation – extends the point to explain how the problem was resolved

Evidence – evidence is given of further inaccuracy

Equipment like flowmeters didn't always work, and gradient was hard to work out using apps on our phones because we didn't want them to get wet. When we did our fieldwork it had rained the day before, so our results would not be reliable compared to a dry summer day. So, in conclusion, we were only partly accurate, and reliability would be difficult depending on the weather.

Judgment – links back to the question, explaining how greater accuracy was achieved

Point – identifies equipment accuracy as a problem

Explanation – explains why reliability is a problem

Judgment – links back to the question, making a judgment

Evidence – an example of reliability

✓ Examiner feedback

See feedback on page 72.

Marked sample answer 2

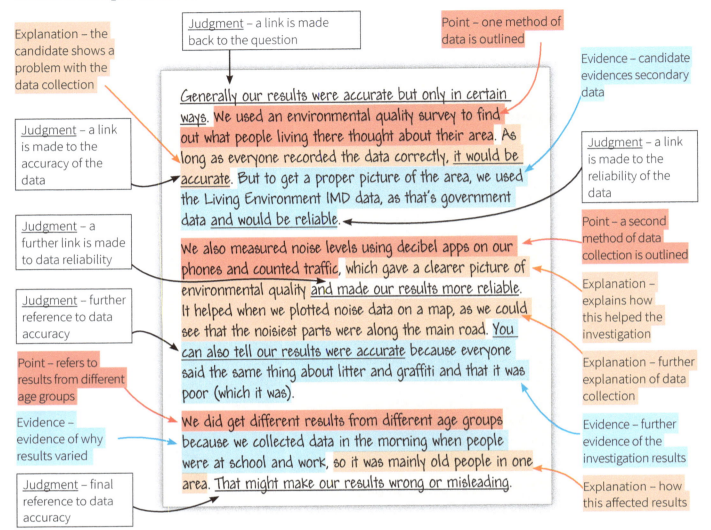

Explanation – the candidate shows a problem with the data collection

Judgment – a link is made back to the question

Point – one method of data is outlined

Evidence – candidate evidences secondary data

Judgment – a link is made to the accuracy of the data

Judgment – a link is made to the reliability of the data

Judgment – a further link is made to data reliability

Point – a second method of data collection is outlined

Judgment – further reference to data accuracy

Explanation – explains how this helped the investigation

Point – refers to results from different age groups

Explanation – further explanation of data collection

Evidence – evidence of why results varied

Evidence – further evidence of the investigation results

Judgment – final reference to data accuracy

Explanation – how this affected results

> Generally our results were accurate but only in certain ways. We used an environmental quality survey to find out what people living there thought about their area. As long as everyone recorded the data correctly, it would be accurate. But to get a proper picture of the area, we used the Living Environment IMD data, as that's government data and would be reliable.
>
> We also measured noise levels using decibel apps on our phones and counted traffic, which gave a clearer picture of environmental quality and made our results more reliable. It helped when we plotted noise data on a map, as we could see that the noisiest parts were along the main road. You can also tell our results were accurate because everyone said the same thing about litter and graffiti and that it was poor (which it was).
>
> We did get different results from different age groups because we collected data in the morning when people were at school and work, so it was mainly old people in one area. That might make our results wrong or misleading.

✓ Examiner feedback

The descriptors for Level 3 applies to both answers as follows:

For AO3:

- Applies understanding to deconstruct information and make logical connections between concepts.
- A balanced, well-developed argument that draws together relevant points coherently.
- Makes judgments supported by evidence.

The candidate has been able to mention both accuracy and reliability in these investigations. Three issues that refer to evidence are explored in both answers, and there are examples of real fieldwork which are explained in detail. Links are made back to the questions and both answers are top quality assessments of each investigation.

For AO4:

- Uses geographical skills to obtain accurate information that supports most of the argument.

This is also strong in both answers. The candidate refers to specific methods of data collection.

By meeting all descriptors fully, both answers are top Level 3 in quality, and are worth all 8 marks.

- In this section you'll learn how to tackle 8-mark questions which use 'Evaluate' as a command word in Paper 1.

Question

Evaluate the evidence that suggests that the global climate is currently changing.   **(8 marks)**

---

 1 Plan your answer

Before attempting to answer the question, remember to **BUG** it. That means:

✔ **Box** the command word.
✔ **Underline** the following:
- the theme
- the focus
- any evidence required
- the number of examples needed.

✔ **Glance** back over the question – to make sure you include everything in your answer.

Use the BUG below to plan your own answer.

Command word: 'Evaluate' means 'judge on its strengths and weaknesses'. You need to decide whether evidence is strong or weak.

Evidence: Use evidence from your own knowledge and understanding, such as shrinking glaciers and seasonal weather changes.

Evaluate the evidence that suggests that the global climate is currently changing. **(8 marks)**

Focus and number of examples: The focus is evidence for a changing global climate. For an 8-mark question, you need three points which are well developed. Each piece of evidence needs to be written in a paragraph. You also need a mini-conclusion.

Theme: Climate change is linked to the theme *Hazardous Earth*, assessed in Paper 1, Section A of your exam. The question is compulsory.

Remember!
- 'Evaluate' is assessed in Paper 1 using your own knowledge and understanding.
- It assesses AO2 (your understanding of a topic) and AO3 (your ability to apply what you know and understand to the question, and make a judgment).
- You must use what you know about climate change to answer this question.

Six steps to success!

The following six activity steps are used in this chapter to help you get the best marks.

1 **Plan your answer** – decide how to structure your answer.

2 **P-E-E-L your answer** – plan what you want to say and the points you want to make.

3 **Write your answer** – use the answer spaces to write your answer.

4 **Mark your answer** – use the mark scheme to self- or peer-mark your answer. You can also use this to mark sample answers in step 5 below.

5 **Mark different answers** – sample answers are given to show you how to mark an answer.

6 **Review the marks for different answers** – these are the same answers as for step 5, but are marked and annotated, so that you can compare these with your own.

2 P-E-E-L your answer

Use PEEL notes to structure your answer. This will help you to communicate your ideas to the examiner in the clearest way. PEEL has four stages:

- **P**oint – You need to make **three** points for an 8-mark answer, based on your understanding of the evidence for climate change, e.g. *'Point 1 – There is a great deal of evidence from recent decades.'*
- **E**vidence – Include details from named examples to support each point, based on your understanding – e.g. *'CO_2 emissions are increasing everywhere in the world.'*
- **E**xplanation – This is your explanation for the evidence, e.g. *'This is because if every CO_2 recording station in the world shows the same pattern then it must be reliable.'* Start your explanation with *'This is because …'*
- **L**ink – Link back to the question about how reliable the evidence is. Finish it off with a one- or two-sentence conclusion about how strong the evidence is that climate is changing.

Tip

Evaluate means stating how strong each piece of evidence is

Don't just describe and explain; show the strength of research behind it. For example, the *changing global climate* might be about shrinking glaciers or increased storminess. You need to say whether this is strong evidence or not. Then draw your argument together in a **mini-conclusion** – it need only be a sentence or two.

Planning grid

Use this planning grid to help you write high-quality paragraphs. Use the guidance in section 6.2 (page 53) to help you.

Note that the fourth row helps you to focus on 'evaluate' – to **evaluate the evidence**.

	Paragraph 1	Paragraph 2	Paragraph 3
Point			
Evidence (*from your own knowledge*)			
Explanation			
Evaluation of the evidence			

✎ **3 Write your answer**

Evaluate the evidence that suggests that the global climate is currently changing. **(8 marks)**

_____

_____

_____

_____

_____

_____

_____

_____

_____

_____

_____

_____

_____

_____

_____

_____

_____

Strengths of the answer	
Ways to improve the answer	

Level		Mark	

4 Mark your answer

1   To help you to identify if the answer includes well-structured points, first highlight or underline:

  • points in red   • explanations in orange   • evidence in blue
  • links to the question that show evaluation.

2   Use the mark scheme below to decide what mark to give. 8-mark questions are not marked using individual points, but instead you should choose a level and a mark based upon the quality of the answer as a whole.

Level	Marks	Descriptor	Examples
	0	No acceptable response	
1	1–3	**For AO2** • Shows isolated elements of understanding of concepts and links between places, environments and processes. **For AO3** • Attempts to apply understanding to deconstruct information but understanding and connections are flawed. • An imbalanced or incomplete argument. • Judgments supported by limited evidence.	• *World temperatures are going up all the time and winters are getting warmer.* • *Global warming is making the seasons different and there are more floods.* • *Scientists think more floods and storms are caused by global warming.*
2	4–6	**For AO2** • Shows elements of understanding of concepts and links between places, environments and processes. **For AO3** • Applies understanding to deconstruct information and give some logical connections between concepts. • An imbalanced argument that draws together some points. • Judgments supported by some evidence.	• *Scientists show that global sea levels have risen in the past 100 years.* • *This is due to global warming which increases temperatures and melts ice caps and glaciers, which go into the sea.* • *We know sea level is rising because countries with coastlines are getting flooded.*
3	7–8	**For AO2** • Shows accurate understanding of concepts and links between places, environments and processes. **For AO3** • Applies understanding to deconstruct information and make logical connections between concepts. • A balanced, well-developed argument that draws together relevant points coherently. • Makes judgments supported by evidence.	• *IPCC research shows that average global sea level has risen by 10–20 cm since 1920.* • *This is probably due to rising global temperatures which melt ice caps, so more water goes into the sea.* • *This is likely to be reliable evidence as the IPCC consists of thousands of the world's best scientists.*

5 Mark different answers

Read through Sample answers 1 and 2.
a) Go through each one using the three colours in section 4, including underlining any evaluative points.
b) Use the level descriptions to decide how many marks it is worth.

Sample answer 1

Many sources of evidence show how climate is changing. Temperatures have risen globally by about 0.8 °C since the 19th century. This is probably due to carbon emissions of greenhouse gases like CO_2 from burning fossil fuels.

Temperatures seem to be getting warmer all the time, so that sea levels will carry on rising. Already some islands in the Pacific have been flooded and countries like Bangladesh have severe floods because much of the country is very low lying. Glaciers in mountains like the Himalayas have been melting because temperatures are rising, so that this all goes to the sea via rivers and makes sea level rise.

Another piece of evidence is that the seasons seem to be changing, so that spring is earlier, and winters are not as cold as they were, and have less snow. Birds migrate earlier than they did and their nests are being built nine days earlier than forty years ago. So that all seems to mean that there is a lot of evidence that climate is changing.

Use a copy of the marking grid below to assess this answer.

Sample answer 2

Globally the climate is warming, and there is evidence to prove that this is the case. Global temperatures are 1 °C warmer than they were 100 years ago because greenhouse gas emissions have increased. It is hard to know exactly what temperatures were like in 1900, and more people and organisations record the weather now than at that time, but there were thermometers, just fewer of them. So, some of the evidence could be questionable, because there were fewer recordings.

Even if temperature recordings are not completely reliable, there is a lot of evidence to show that sea level has risen globally by about 20 cm in 100 years, partly because ocean water expands when it warms and so it rises. Many coastal areas are flooding more now, so it is a global process and not just evidence from one place.

Other evidence that shows temperatures are rising comes from retreating glaciers and ice sheets because they are melting. Many glaciers have been photographed for over 100 years, and many in the Alps and on Greenland show that they have retreated a long way from where they once were.

Strengths of the answer		
Ways to improve the answer		
Answer Level		Mark

6 Review the marks for different answers

Sample answers 1 and 2 are marked below. The following have been highlighted to show how well each answer has structured points:

- points in red   • explanations in orange   • evidence in blue
- evaluations are underlined. These are important to reach Level 3 on questions whose command word is 'Evaluate'.

Marked sample answer 1

Many sources of evidence show how climate is changing. Temperatures have risen globally by about 0.8 °C since the 19th century. This is probably due to carbon emissions of greenhouse gases like CO_2 from burning fossil fuels.

Temperatures seem to be getting warmer all the time, so that sea levels will carry on rising. Already some islands in the Pacific have been flooded and countries like Bangladesh have severe floods because much of the country is very low lying. Glaciers in mountains like the Himalayas have been melting because temperatures are rising, so that this all goes to the sea via rivers and makes sea level rise.

Another piece of evidence is that the seasons seem to be changing, so that spring is earlier, and winters are not as cold as they were, and have less snow. Birds migrate earlier than they did and their nests are being built nine days earlier than forty years ago. So that all seems to mean that there is a lot of evidence that climate is changing.

Point – quantifying the amount of warming

Explanation – a reason is given for warming of the global climate

Point – the candidate offers further evidence

Explanation – a reason is given for flooding in many countries

Point – further evidence is given for climate change

Evidence – the candidate discusses retreating glaciers as evidence for rising sea levels

Point – further evidence; winters are warmer now

Explanation – a reason is given for glaciers melting

Evidence – the point is extended with the example of bird migrations

Examiner feedback

Examiners often see this kind of answer. This candidate knows a lot and has learned facts and figures. The answer is a problem though, because there is no evaluation. The candidate needs to ask themselves – *'what's the evidence that glaciers are melting, and is it reliable? How do I know it's reliable?'*.

The answer is therefore a mix of levels:

- Almost Level 3 for AO2, understanding climate change and global warming.
- Explanations are mid-Level 2 because they do not always link to warming climate (e.g. flooding in Bangladesh is explained because it is low lying, not because of sea level change).
- However, there is no evaluation, an essential quality for high Level 2 or Level 3.

Faced with this, examiners have to do a 'best fit' or a kind of average. The examiner gives this low Level 2 and 4 marks.

Marked sample answer 2

Explanation – the candidate briefly explains the increase in temperatures

Point – the candidate makes the point about increasing temperatures

Evaluation – one reason given why temperature readings may not be accurate

Evaluation – the evaluation is extended by referring to volume of temperature recordings

Point – a second point about rising sea level

Explanation – the candidate gives a reason for this

Evaluation – the candidate shows that this is probably reliable as many places experience the same thing

Point – a point about retreating glaciers

Evaluation – reference to the reliability of photos taken over a long time to show change

Explanation – the candidate explains the point about retreating glaciers

Globally the climate is warming, and there is evidence to prove that this is the case. Global temperatures are 1 °C warmer than they were 100 years ago because greenhouse gas emissions have increased. It is hard to know exactly what temperatures were like in 1900, and more people and organisations record the weather now than at that time, but there were thermometers, just fewer of them. So, some of the evidence could be questionable, because there were fewer recordings.

Even if temperature recordings are not completely reliable, there is a lot of evidence to show that sea level has risen globally by about 20 cm in 100 years, partly because ocean water expands when it warms and so it rises. Many coastal areas are flooding more now, so it is a global process and not just evidence from one place.

Other evidence that shows temperatures are rising comes from retreating glaciers and ice sheets because they are melting. Many glaciers have been photographed for over 100 years, and many in the Alps and on Greenland show that they have retreated a long way from where they once were.

 Examiner feedback

This is a top quality answer which was given the full 8 marks.

Notice that the candidate giving this answer has shown less knowledge and understanding (AO2) than the candidate giving Sample answer 1, but nearly half of the answer is spent showing whether the evidence for change is reliable or not (AO3). That's what you need to do in a question whose command word is 'Evaluate'. Spend as much time on evaluating as you do in showing your knowledge and understanding.

On your marks!

6.6 8-mark questions using 'Evaluate' in Paper 2

- **In this section you'll learn how to tackle 8-mark questions which use 'Evaluate' as a command word in Paper 2.**

Study **Figure 1**, a photo showing deposition of sediment along a stretch of coast in South Australia.

Figure 1

Question

Using **Figure 1**, evaluate the part played by sediment deposition in creating coastal landscapes.   **(8 marks plus 4 marks SPaG)**

How is SPaG assessed?

One 8-mark question on each of Papers 1 and 2 will assess spelling, punctuation, grammar and the use of specialist terminology (SPaG); 4 marks are allocated as follows:

- high performance (4 marks)
- intermediate performance (2–3 marks)
- threshold performance (1 mark).

Examiners mark SPaG based on your:

- spelling accuracy, including capitalisation
- punctuation – the use of commas, full stops and semi-colons. Try reading an answer aloud; if it leaves you gasping for breath, it needs more punctuation!
- syntax – i.e. the quality of your grammar
- use of paragraphs.

Six steps to success!

The following six activity steps are used in this chapter to help you get the best marks.

1 **Plan your answer** – decide how to structure your answer.

2 **P-E-E-L your answer** – plan what you want to say and the points you want to make.

3 **Write your answer** – use the answer spaces to write your answer.

4 **Mark your answer** – use the mark scheme to self- or peer-mark your answers. You can also use this to mark a sample answer in step 5 below.

5 **Mark a different answer** – a sample answer is given to show you how to mark an answer.

6 **Review the marks for a different answer** – this is the same answer as for step 5, but is marked and annotated, so that you can compare this with your own.

Level	Marks	Descriptor
	0	Writes nothing or in a style which does not link to the question, or make sense of the question.
1	1	Spelling and punctuation reasonably accurate. Some meaning overall. A limited range of specialist terms.
2	2–3	Spelling and punctuation show considerable accuracy. Grammar shows general control of meaning overall with a good range of specialist terms.
3	4	Spelling and punctuation show consistent accuracy. Grammar shows effective control of meaning overall with a wide range of specialist terms.

Figure 2 *Mark scheme for SPaG*

1 Plan your answer

Before attempting to answer the question, remember to **BUG** it using the guidelines on pages 34–35.

Annotate the question in the space below.

> Using **Figure 1**, evaluate the part played by sediment deposition in creating coastal landscapes. **(8 marks, plus 4 marks SPaG)**

Remember!

- 'Evaluate' is assessed in Paper 2 using a resource stimulus, like the photo in this example.
- It assesses AO4 (your skill in interpreting the photo) and AO3 (your ability to apply what you see to the question).
- You can use what you know about coastal deposition as well as what's in the photo.

2 P-E-E-L your answer

Use PEEL notes to structure your answer. This will help you to communicate your ideas to the examiner in the clearest way. PEEL has four stages:

- **P**oint – You need to make **three** points for an 8-mark answer, based on your understanding of sediment deposition, e.g. *'Point 1 – Sediment deposition is important along coasts where longshore drift occurs.'*
- **E**vidence – Include details from the photo to support each point, e.g. *'The photo shows a spit which consists of sand transported there by longshore drift.'*
- **E**xplanation – This is your explanation for the evidence, e.g. *'This is because the spit has extended along the shore, even diverting a river as it has grown.'* Start your explanation with *'This is because …'*
- **L**ink – Link back to the question about ways in which sediment contributes to coastal landscapes. Finish it off with a one- or two-sentence conclusion about the contribution that sediment deposition can make.

Tip

Evaluate means stating how strong each piece of evidence is.

Don't just describe and explain: show the strength of evidence. For example, if a coastal spit is the most significant feature along a stretch of coast, then sediment deposition makes a big contribution. Draw your argument together in a **mini-conclusion** – it need only be a sentence or two.

Planning grid

Use this planning grid to help you write high-quality paragraphs. Use the guidance in section 6.2 (page 53) to help you. Note that this is an 8-mark question, so needs three PEEL Paragraphs.

Note that the fourth row helps you to focus on 'evaluate' to help you link back to the question – to **evaluate the evidence**.

	Paragraph 1	Paragraph 2	Paragraph 3
Point			
Evidence (*from your own knowledge*)			
Explanation			
Link – evaluate the evidence			

3 Write your answer

Using **Figure 1**, evaluate the part played by sediment deposition in creating coastal landscapes.

(8 marks, plus 4 marks SPaG)

_____

_____

_____

_____

_____

_____

_____

_____

_____

_____

_____

_____

_____

_____

_____

_____

Strengths of the answer	
Ways to improve the answer	

Level		Mark	

✎ **4 Mark your answer**

1   To help you to identify if the answer includes well-structured points, first highlight or underline:

- points in red   • explanations in orange   • evidence in blue
- links to the question that show evaluation.

2   Use the mark scheme below to decide what mark to give. 8-mark questions are not marked using individual points, but instead you should choose a level and a mark based upon the quality of the answer as a whole.

3   Remember to give a mark for SPaG!

Level	Marks	Descriptor	Examples
	0	No acceptable response	
1	1–3	**For AO3** • Attempts to apply understanding to deconstruct information but understanding and connections are flawed. • An imbalanced or incomplete argument. • Judgments supported by limited evidence. **For AO4** • Uses some geographical skills to obtain information with limited relevance and accuracy, which supports few aspects of the argument.	• *The spit comes from waves which break on the beach and longshore drift takes place.* • *The photo shows a sandy beach which reaches almost across the river.* • *Deposition creates many features like spits and beaches.*
2	4–6	**For AO3** • Applies understanding to deconstruct information and give some logical connections between concepts. • An imbalanced argument that draws together some points. • Judgments supported by some evidence. **For AO4** • Uses geographical skills to obtain accurate information that supports some aspects of the argument.	• *Coastal spits are formed when waves break on the shore at an angle and take sediment along the coast forming a long, sandy headland into the water.* • *Figure 1 shows how the river stops the spit from forming a bar which would join the two bits of coast together. So deposition can create important landforms.* • *There might be many coastal deposits along a coast, like sandy beaches which lead to resorts like Brighton.*
3	7–8	**For AO3** • Applies understanding to deconstruct information and make logical connections between concepts. • A balanced, well-developed argument that draws together relevant points coherently. • Makes judgments supported by evidence. **For AO4** • Uses geographical skills to obtain accurate information that supports all aspects of the argument.	• *The coastal spit shown has been formed by two sets of processes. The main one is longshore drift, caused by winds creating waves which hit the shore at an angle.* • *Figure 1 shows a coastal spit which has forced the river to divert from where it used to reach the sea. This shows the impact of sediment – it can divert features such as rivers.* • *Spits are really significant landforms, like Spurn Head in East Yorkshire which shelters the Humber from storms.*

5 Mark a different answer

Read through the sample answer below.

a) Go through it using the three colours in section 4, including underlining any evaluative points.

b) Use the level descriptions to decide how many marks it is worth.

The photo shows a spit formed of sand that has been deposited on the beach. The waves approach at an angle and swash takes the sand up the beach, where it runs back down in a zig-zag pattern. Further waves repeat the process, so an elongated spit is formed. The spit in the photo moved until it reached the river, and the river current has then shaped it where it runs out to sea. The river in the photo has been diverted around the spit. Deposition can therefore divert river flow, showing how important it is.

Another depositional landform is a sand bar, which is like a spit except that there is no river to prevent movement of sand. The bar develops until it cuts off a lagoon. The lagoon can create important wildlife refuges because freshwater behind the bar remains sheltered and ideal for wildfowl, especially in winter. This means that deposition can create important features of coastal landscapes, like mud flats behind a spit, which are areas of calm water away from storms.

The final contribution made by coastal deposition is beaches, which have physical impacts because they protect cliffs from erosion by absorbing friction from advancing waves. Where longshore drift moves beach material away, it may increase coastal retreat.

Strengths of the answer			
Ways to improve the answer			
Level		**Mark out of 8**	
SPaG level		**Mark out of 4**	

6 Review the marks for a different answer

The sample answer is marked on the next page. The following have been highlighted to show how well it has structured points:

- points in red
- explanations in orange
- evidence in blue
- links to the question that show evaluation are underlined.

Explanation – the process is described by which the spit forms

Point – identifies a spit and makes it clear it's depositional

Evidence – the candidate evidences the process from the photo. This kind of evidence is important when you need to explain processes as a sequence of stages

The photo shows a spit formed of sand that has been deposited on the beach. The waves approach at an angle and swash takes the sand up the beach, where it runs back down in a zig-zag pattern. Further waves repeat the process, so an elongated spit is formed. The spit in the photo moves until it reached the river, and the river current has then shaped it where it runs out to sea. The river in the photo has been diverted around the spit. Deposition can therefore divert river flow, showing how important it is.

Another depositional landform is a sand bar, which is like a spit except that there is no river to prevent movement of sand. The bar develops until it cuts off a lagoon. The lagoon can create important wildlife refuges because freshwater behind the bar remains sheltered and ideal for wildfowl, especially in winter. This means that deposition can create important features of coastal landscapes, like mud flats behind a spit, which are areas of calm water away from storms.

The final contribution made by coastal deposition is beaches, which have physical impacts because they protect cliffs from erosion by absorbing friction from advancing waves. Where longshore drift moves beach material away, it may increase coastal retreat.

Link – shows the importance of depositional features (evaluation)

Point – identifies a sand bar and makes it clear it's depositional

Explanation – the process of bar formation is explained

Link – uses wording of the question about the importance of features of landscapes (evaluation)

Point – beaches are named as a third depositional landform

Evidence – exemplifies mud flats as important features

Explanation – the importance of beaches is explained

 Examiner feedback

This is a good answer. This candidate explains three landforms, and uses the photo to identify features. The photo is used in the first paragraph, but not afterwards, so AO4 is less here than AO3, where the candidate shows some good explanation. The answer needs a final mini-conclusion in the last sentence.

The descriptors for Level 3 for AO2 applies to this answer:

- *'Applies understanding to deconstruct information and make logical connections between concepts'* – the candidate shows an ability to describe landform formation in some detail. Landforms are identified and processes described.
- *'A balanced, well-developed argument that draws together relevant points coherently'* – the candidate develops an argument to show how important deposition is.
- *'Makes judgments supported by evidence'* – the candidate refers to the photo meaningfully, firstly by naming the landforms, and secondly by explaining the significance of the landforms.

Because the photo is not referred to much in the answer, the descriptor for Level 2 for AO4 applies:

- *Uses geographical skills to obtain accurate information that supports some aspects of the argument* – more use could be made of the photo.

The mark given is therefore a best fit between AO3 and AO4, so the candidate is given 6 marks.

For SPaG, the answer is given 4 marks – spelling, syntax, and paragraphing are all good.

On your marks!

6.7 Hitting the high marks on 12-mark questions in Paper 3

- In this section you will learn how to prepare for the 12-mark questions in Paper 3, which use the command word 'Justify'.

What's different about Paper 3?

Paper 3 is different from Papers 1 and 2.

- It is a decision-making exercise (DME) about a geographical issue assessing the biosphere (Topic 7) and resources (Topic 9), focusing on a place which will be located in either a taiga or tropical rainforest region (Topic 8).
- The issue will be given to you in a Resource Booklet (of about 10 pages). It will probably be about a place that you haven't studied. Don't worry about this – it is your ability to read and interpret the booklet that is being assessed, not your knowledge of the place.
- The exam lasts for 1 hour 30 minutes. It has a total of 64 marks, including 4 marks for spelling, punctuation, grammar and use of specialist terminology (SPaG). This means it's less pressured for time than the other papers.
- The last question will ask you to make a decision about the issue in the Resource Booklet. That will need thinking and planning time – that's why the timing of the paper is different.

Spending your time in the exam

- It's essential that you read the Resource Booklet carefully, so that you become familiar with the information.
- The Resource Booklet will give you information about the issue and the place, and will lead up to proposals for the future. It will contain text, maps, photographs, graphs, tables of data, and views and opinions about the issue.
- Use these resources to help you make sense of the place and the issue.
- Most exam questions will test your understanding of the booklet, though some will test your wider understanding of Topics 7, 8 and 9.
- Some of the questions will involve making calculations, so you'll need to have access to a calculator.

Preparing for Paper 3

Both the Resource Booklet and exam paper are organised into sections:

- Section A: People and the biosphere (Topic 7)
- Section B: Forests under threat (Topic 8)
- Section C: Consuming energy resources (Topic 9)
- Section D: Making a geographical decision (Topics 7, 8 and 9)

To prepare for the exam, you will need to revise all three topics, because some shorter questions will test your knowledge and understanding of them (AO2). You will also need to apply your knowledge and understanding (AO3) along with interpreting information in the Resource Booklet (AO4).

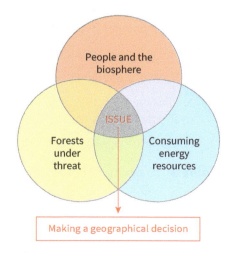

Figure 1 *How Topics 7, 8 and 9 link together*

Recipes for success in Paper 3

The exam paper will always have the same format (Figure 2).

Paper 3: **People and environment** **issues – Making a** **geographical decision** You'll be given a Resource Booklet Total marks: **64** *including 4 marks for SPaG* Weighting: **25%** Time: **1 hr 30 mins**	**Section A: People and the biosphere** • Marked out of 8. • Answer all questions in Section A.
	Section B: Forests under threat • Marked out of 7. • Answer all questions in Section B.
	Section C: Consuming energy resources • Marked out of 33, including two 8-mark questions. • Answer all questions in Section C.
	Section D: Making a geographical decision • The decision will be about an energy issue in either tropical rainforest or taiga. • One 12-mark question plus 4 marks SPaG (16 marks).

Figure 2 *The format of Paper 3*

The questions in the examination follow a decision-making process (Figure 3). They're not separate, but linked into a sequence.

- The issue is stated at the start of the Resource Booklet.
- Read the booklet first and then think about the issue.
- The most demanding question is in Section D, where you'll have to justify a choice from three options.
- There is no preferred option among examiners. All options can be justified. Your reasons for making a choice will vary according to the chosen option.
- It's marked by assessing a mix of AO2 (understanding from the course), AO3 (applying and developing your argument), and AO4 (skill in selecting information from the Resource Booklet).

Stage		Where you can find this in the exam
1 Identify a problem or a need.		Section A
2 Exploring the reasons for the problem.		Section B
3 Look at the issues behind the problem.		Section C
4 Identify different solutions/options and weigh up advantages/disadvantages.		Section D
5 Make a decision: which option is best?		Section D

Figure 3 *The decision-making process*

How to justify an argument in Section D

To do well on the 12-mark question, your answer will need to:

- make a clear choice of one option
- explain its impact on the economy, people and environment
- weigh up advantages/disadvantages of all three options
- make an overall judgment.

To make a choice, consider the following:

Economic factors

- Will the number of jobs increase, including those which are higher skilled and higher paid?
- Will there be an increase in GDP?
- Will the area become more attractive for investment?

Social factors

- Will people's quality of life improve?
- Will there be better housing, health and education?

Environmental factors

- Will air and water quality improve/reduce pollution?
- Will wildlife be protected/conserved?
- Will the built environment be improved?

Checklist

- Have you written about all three options?
- Have you given a clear argument justifying your choice of the best option? You need at least two advantages, and two disadvantages explaining why you rejected the other two options. (AO3)
- Have you given detailed evidence from the booklet? (AO4)
- Have you included some knowledge and understanding from Topic 7 (People and the biosphere) to Topic 9 (Consuming energy resources)? (AO2)

Writing a top quality answer

You are going to assess a sample answer to the question below (it is question 4 on page 138 of this book).

Study the **three** options below for how DRC should provide sufficient energy to provide for its population and to further develop the country in the 21st century. **(12 marks and SPaG 4 marks)**

- **Option 1**: Cease building large-scale, top-down, energy generating schemes and encourage more small-scale, bottom-up projects.
- **Option 2**: Adopt more sustainable management of DRC's rainforests including replanting to replace deforestation caused by building energy installations.
- **Option 3**: Expand large-scale HEP generation as the only practical way of solving DRC's energy needs.

Select the option that would be the best plan to ensure **long-term** energy security for the economy, and the environment.

Read the following answer to the question. Using highlighters, pick out where the candidate has:

- put together an argument (AO3)
- picked out detail from the Resource Booklet (AO4)
- used information that isn't in the booklet but the candidate has learned from the rest of the course (AO2).

In pairs, come to a mark out of 12 and decide on a mark out of 4 for SPaG. Check your verdict with the marked example on the next page.

The DRC government faces many challenges. It needs energy to develop, but has the problem of severe environmental impacts. DRC's population is growing fast (Figure 6), probably because death rates are falling. It needs to grow economically to sustain its population, to grow crops to feed people, and also for employment. Option 1 can provide these, but in ways that avoid large-scale top-down development.

Option 1 is most suitable as it allows small-scale economic development and therefore helps to protect remaining rainforests. This is because the government would be able to attract new industries with existing energy resources, and also benefit from the economic multiplier effect of new jobs, and from the REDD scheme attracting money from Norway because DRC would be protecting its rainforests from further destruction. This would be needed, because large-scale rainforest destruction through the Inga dams (Figure 10) is enormous, and this damage needs to be prevented.

Socially, Option 1 would benefit forest indigenous peoples. Figure 7 shows the benefits that people gain from the rainforest, e.g. food, medicines. Their lives depend on the forest, from crops to hunting and gathering. This has the potential for ecotourism, as does Option 2, as people would want to see traditional lifestyles, which are disappearing across the world as rainforests are destroyed. Tourist development would also increase the economic multiplier and also at a small, local scale.

I rejected Option 2, though it came close second. Rainforests should be replanted, but the only way DRC will develop in future is if energy resources are plentiful. That's the benefit of Option 1. Some African countries are starting to develop economically because labour costs are cheap compared to other countries, and they need to develop economically so that they gain through jobs. Option 2 feels like development would be small scale.

I rejected Option 3 because it would destroy rainforest, and DRC would receive no income from Norway for maintaining forests. Clearance would also remove carbon sinks and increase the percentage of greenhouse gases in the atmosphere, leading to faster climate change. It would acidify water in the dams as rainforests rotted below water level. It seems to have no advantages.

To conclude, Option 1 is best for where DRC is now. Energy is needed but not as Option 3 suggests.

Marked example

Read the annotations below.

- Introduction and conclusions are <u>underlined.</u>
- Parts of the argument (AO3) are shown in red.
- Details from the Resource Booklet (AO4) are shown in blue.
- information learned from the rest of the course (AO2) is in orange.

<u>The DRC government faces many challenges. It needs energy to develop, but has the problem of severe environmental impacts.</u> DRC's population is growing fast (Figure 6), probably because death rates are falling. <u>It needs to grow economically to sustain its population, to grow crops to feed people, and also for employment. Option 1 can provide these, but in ways that avoid large-scale top-down development.</u>

Option 1 is most suitable as it allows small-scale economic development and therefore helps to protect remaining rainforests. This is because the government would be able to attract new industries with existing energy resources, and also benefit from the economic multiplier effect of new jobs, and from the REDD scheme attracting money from Norway because DRC would be protecting its rainforests from further destruction. This would be needed, because large-scale rainforest destruction through the Inga dams (Figure 10) is enormous, and this damage needs to be prevented.

Socially, Option 1 would benefit forest indigenous peoples. Figure 7 shows the benefits that people gain from the rainforest, e.g. food, medicines. Their lives depend on the forest, from crops to hunting and gathering. This has the potential for ecotourism, as does Option 2, as people would want to see traditional lifestyles, which are disappearing across the world as rainforests are destroyed. Tourist development would also increase the economic multiplier and also at a small, local scale.

I rejected Option 2, though it came close second. Rainforests should be replanted, but the only way DRC will develop in future is if energy resources are plentiful. That's the benefit of Option 1. Some African countries are starting to develop economically because labour costs are cheap compared to other countries, and they need to develop economically so that they gain through jobs. Option 2 feels like development would be small scale.

I rejected Option 3 because it would destroy rainforest, and DRC would receive no income from Norway for maintaining forests. Clearance would also remove carbon sinks and increase the percentage of greenhouse gases in the atmosphere, leading to faster climate change. It would acidify water in the dams as rainforests rotted below water level. It seems to have no advantages.

<u>To conclude, Option 1 is best for where DRC is now. Energy is needed but not as Option 3 suggests.</u>

 Examiner decision

This answer gains the full 12 marks, plus 4 marks for SPaG. It meets AO2, AO3 and AO4 well, and the quality of SPaG is very high.

GCSE 9-1 Geography Edexcel B
Practice Paper 1

Global Geographical Issues

Time allowed: 1 hour 30 minutes
Total number of marks: 94 (including 4 marks for spelling,
punctuation, grammar and use of specialist terminology (SPaG))

Instructions
Answer **all** questions.

Answer ALL questions. Write your answers in the spaces provided.

Some questions must be answered with a cross in a box ☒. If you change your mind about an answer, put a line through the box ☒ and then mark your new answer with a cross ☒.

1 (a) (i) Identify **one** way in which tectonic plates move at a conservative boundary.

(1)

☒ **A** Towards one another
☒ **B** Side by side
☒ **C** Away from each other
☒ **D** There is no movement

(ii) Identify which **one** of the following locations is on a conservative plate boundary.

(1)

☒ **A** Iceland
☒ **B** California
☒ **C** Hawaii
☒ **D** Western coast of South America

(b) Explain **one** reason why the majority of volcanoes and earthquakes occur at plate boundaries.

(2)

_____

_____

_____

_____

(c) (i) A total of 1482 earthquakes occurred in a seven-day period at the end of April 2016.
Calculate the mean number of earthquakes per day, to one decimal place. Show your working.

(2)

Answer = _____

(ii) Explain **one** secondary effect of **either** an earthquake **or** a volcano.

Chosen event (please circle one): **earthquake / volcano**

(2)

_____

_____

_____

_____

(d)   Study **Figure 1**. It shows a diagram of the global atmospheric circulation.

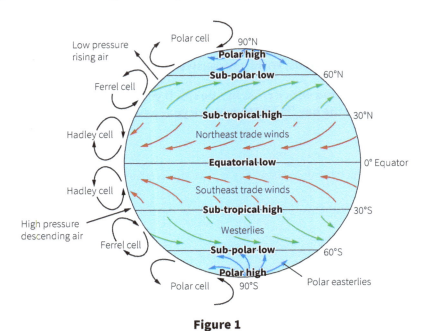

Figure 1

(i)   Using **Figure 1**, identify which **two** of the following statements are true.

(2)

☒ **A**   Winds blow from low to high pressure.
☒ **B**   High pressure is an area of sinking air.
☒ **C**   The south-east trades blow in the northern hemisphere.
☒ **D**   Surface winds are named after the direction that they are blowing towards.
☒ **E**   The UK lies in the westerlies wind belt.
☒ **F**   Sinking air above the Equator forms the Hadley Cell.

(ii)   State **two** effects of the Earth's revolution around the Sun on the pressure and
wind belts in **Figure 1**.

(2)

1 _____

2 _____

(iii) Explain **two** ways in which the global atmospheric circulation determines the location of
the arid regions of the world.

(4)

1 _____

_____

_____

_____

2 _____

_____

_____

(e)   **Figure 2** shows changes in the strength of Atlantic tropical storms between 1950 and 2010.

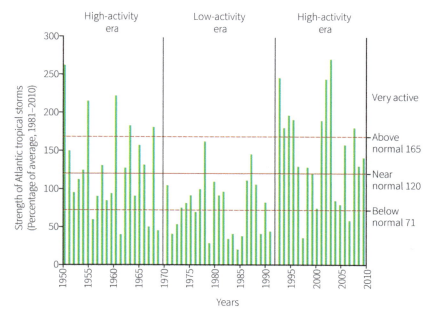

Figure 2

(i)   Identify which **one** of the following data presentation techniques is used in **Figure 2**.

(1)

⊠ **A**   Bar chart
⊠ **B**   Pie chart
⊠ **C**   Line chart
⊠ **D**   Proportional bar graph

(ii)   Suggest **one** reason why there has been an increase in the frequency of strong tropical storms in recent years.

(2)

_____

_____

_____

_____

(iii)   Identify which **one** of the following is the name given to tropical storms in the Atlantic.

(1)

⊠ **A**   Cyclones
⊠ **B**   Hurricanes
⊠ **C**   Typhoons
⊠ **D**   Tsunami

(iv) Explain **one** reason why most tropical storms develop between 5° and 15° north and south of the Equator.

(2)

_____

_____

_____

_____

(f) Evaluate the view that 'the world's developed countries are more effective than developing and emerging countries in their planning and preparation for tropical storms'.

(8)

_____

_____

_____

_____

_____

_____

_____

_____

_____

_____

_____

_____

_____

_____

_____

_____

(Total for Question 1 = 30 marks)
TOTAL FOR SECTION A = 30 MARKS

Answer ALL questions. Write your answers in the spaces provided.

*Spelling, punctuation, grammar and the use of specialist terminology will be assessed in (f).

2 (a) (i) Identify which **one** of the following statements is a part of Frank's 'dependency theory'.

(1)

⊠ **A** A country will go through five stages of development.
⊠ **B** The world can be divided into a core and a periphery.
⊠ **C** The periphery will export high value goods to the core.
⊠ **D** A country's development is not influenced by its history.

(ii) Identify which **one** of the following statements is generally true about a developed country.

(1)

⊠ **A** It will have a population pyramid with a wide base.
⊠ **B** The Total Fertility Rate will be high.
⊠ **C** The literacy rate will be high.
⊠ **D** The number of people per doctor will be high.

(b) **Figure 3** shows global variations in HDI.

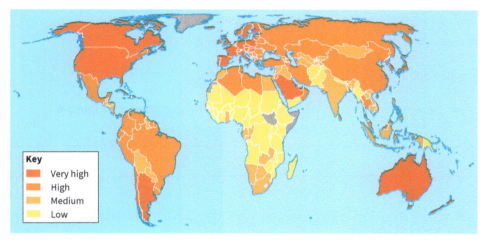

Figure 3

(i) State the meaning of the letters 'HDI'.

(1)

_____

(ii) Explain **two** advantages of using HDI as a development indicator.

(4)

1 _____

_____

_____

_____

2 _____

_____

_____

_____

(c) **Figure 4** is a scattergraph showing the relationship between birth rate and infant mortality rate in a number of countries.

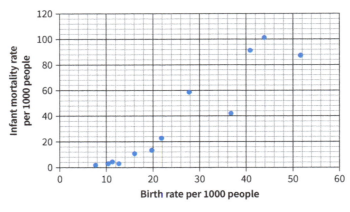

Figure 4

(i)   Plot the following data for the Côte d'Ivoire on **Figure 4**.

Birth rate 45 per 1000   Infant mortality rate 82 per 1000

(1)

(ii) Draw in the best fit line on **Figure 4**.

(1)

(iii) Identify the relationship between infant mortality and birth rate shown on **Figure 4**.

(1)

_____

_____

(d) **Figure 5** shows the Kariba Dam in Zambia which was built to provide hydro-electricity.

Figure 5

Suggest **two** ways in which dams such as the Kariba Dam are examples of 'top-down' development.

(4)

1 _____

_____

_____

_____

2 _____

_____

_____

_____

(e)   **Figure 6** shows an example of intermediate technology used in pumping water.

Figure 6

(i)   Explain **two** ways in which the example in **Figure 6** is an example of intermediate technology.

(4)

1 _____

_____

_____

_____

2 _____

_____

_____

_____

(ii) Explain **one** advantage of intermediate technology as an approach to development.

(2)

_____

_____

_____

_____

(iii) Explain **one** disadvantage of intermediate technology as an approach to development.

(2)

_____

_____

_____

_____

*(f)   In this question, an additional 4 marks will be for spelling, punctuation, grammar, and your use of specialist terminology.

Assess the environmental impacts of economic development in a **named** emerging country.

(8 and 4)

Name of chosen country: _____

_____

_____

_____

_____

_____

_____

_____

_____

_____

_____

_____

_____

_____

_____

(Total for Question 2 = 34 marks)
TOTAL FOR SECTION B = 34 MARKS

Answer ALL questions.

3 (a)   Identify which **one** of the following is the correct definition of the term 'urbanisation'.

(1)

☒ **A**   Unplanned growth into the surrounding urban area.
☒ **B**   When an increasing percentage of a country's population live in towns and cities.
☒ **C**   The process by which suburbs grow as a city grows outwards.
☒ **D**   The redevelopment of former industrial areas or housing to improve them.

(b)   Identify which **one** of the following techniques would be the best way of showing inequalities between different parts of a city.

(1)

☒ **A**   A choropleth map
☒ **B**   A flow line map
☒ **C**   A line chart
☒ **D**   A pie chart

(c)   **Figure 7** gives information about the world's megacities.

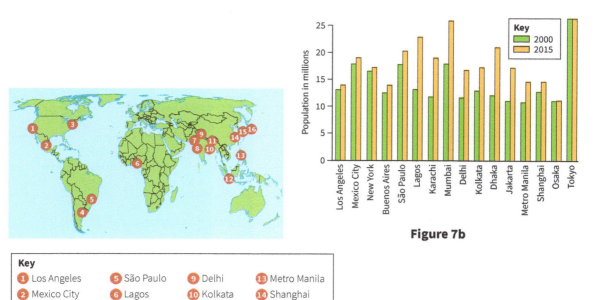

Figure 7b

Key
1 Los Angeles **5** São Paulo **9** Delhi **13** Metro Manila
2 Mexico City **6** Lagos **10** Kolkata **14** Shanghai
3 New York **7** Karachi **11** Dhaka **15** Osaka
4 Buenos Aires **8** Mumbai **12** Jakarta **16** Tokyo

Figure 7a

(i)   Define the term 'megacity'.

(1)

_____

_____

(ii)   Study **Figure 7a**. Describe the **distribution** of the world's megacities.

(2)

_____

_____

_____

_____

(iii) **Study Figure 7b**. Compare the **growth** of the world's megacities between 2000 and 2015.

(3)

_____

_____

_____

_____

_____

_____

(iv)   Explain why some urban areas dominate the rest of the country in which they are located.

(3)

_____

_____

_____

_____

_____

_____

_____

(d)   Name a megacity in **either** a developing **or** an emerging country that you have studied.

Name of chosen megacity: _____

(i)   Describe the **site** of your chosen megacity.

(2)

_____

_____

_____

_____

(ii) Describe the **situation** of your chosen megacity.

(2)

_____

_____

_____

_____

(iii) Explain **two** reasons why your chosen megacity has experienced rapid population growth.

(4)

1 _____

_____

_____

_____

2 _____

_____

_____

_____

(iv) Explain **one** challenge of living in your chosen megacity that is caused by its rapid population growth.

(3)

_____

_____

_____

_____

_____

_____

(e)   Assess the challenges caused by extremes of wealth within your chosen megacity.

(8)

_____

_____

_____

_____

_____

_____

_____

_____

_____

_____

_____

_____

_____

_____

_____

_____

_____

_____

_____

_____

_____

(Total for Question 3 = 30 marks)
TOTAL FOR SECTION C = 30 MARKS
TOTAL FOR THE PAPER = 94 MARKS

GCSE 9-1 Geography Edexcel B
Practice Paper 2

UK Geographical Issues

Time allowed: 1 hour 30 minutes
Total number of marks: 94 (including 4 marks for spelling, punctuation, grammar and use of specialist terminology (SPaG))

Instructions
Answer **all** questions in Section A and Section B
Answer **two** questions in Section C

Answer ALL questions. Write your answers in the spaces provided.

Some questions must be answered with a cross in a box ☒. If you change your mind about an answer, put a line through the box ☒ and mark your new answer with a cross ☒.

***Spelling, punctuation, grammar and the use of specialist terminology will be assessed in question 7.**

1   (a)   **Figure 1** shows the distribution of one major rock type found in the British Isles.

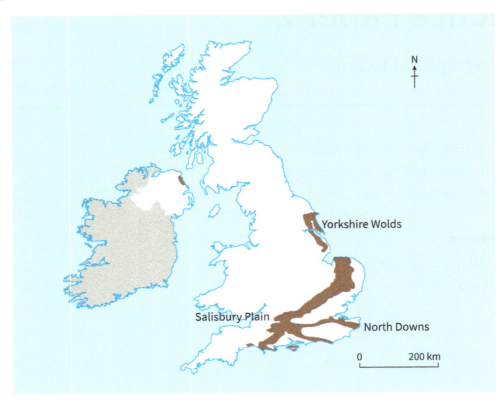

Figure 1

(i)   Identify which **one** of the following rock types is found in the shaded areas in **Figure 1**.

(1)

☒ **A** Carboniferous limestone
☒ **B** Granite
☒ **C** Sandstone
☒ **D** Chalk

(ii) **Figure 2** shows part of an upland part of the United Kingdom.

Figure 2

Identify which **one** type of landscape is shown in **Figure 2**.

(1)

⊠ **A** Tectonic activity
⊠ **B** Glacial upland
⊠ **C** Chalk scarp and clay vale
⊠ **D** The upper course of a river valley

(b) Explain **one** way in which human activity has created distinctive landscapes in the UK.

(2)

_____

_____

_____

_____

(Total for Question 1 = 4 marks)

Coastal Change and Conflict

2 (a) **Figure 3** shows some coastal management techniques.

Figure 3

(i) Identify which **one** coastal management technique is labelled **X** in **Figure 3**.

(1)

☒ **A** Beach nourishment
☒ **B** Rock armour
☒ **C** Groyne
☒ **D** Sea wall

(ii) Identify which **one** coastal management technique is labelled **Y** in **Figure 3**.

(1)

☒ **A** Beach nourishment
☒ **B** Rock armour
☒ **C** Groyne
☒ **D** Sea wall

(b) (i) Using a labelled diagram, explain the process of longshore drift.

(4)

(ii) Explain **one** reason why people in a coastal area may wish to manage the effects of longshore drift.

(2)

_____

_____

_____

_____

(Total for Question 2 = 8 marks)

River Processes and Pressures

3 (a) Identify which of the following is a description of the river process of attrition.

(1)

☒ **A** The scratching and scraping of a river bed and banks by the stones and sand.
☒ **B** The wearing away of particles by the action of other particles in the river's load.
☒ **C** The velocity of a river.
☒ **D** The dissolving of chemicals by the water in the river.

(b) **Figure 4** shows a storm hydrograph.

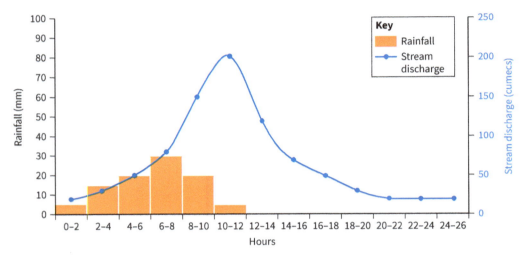

Figure 4

(i) Define the term 'stream discharge'.

(1)

_____

_____

(ii) Identify the peak discharge in cumecs.

(1)

Answer = _____ cumecs

(iii) Explain **two** ways in which land-use changes can increase the risk of a river flooding.

(4)

1 _____

_____

_____

_____

2 _____

_____

_____

_____

(Total for Question 3 = 7 marks)

Investigating a Geographical Issue

4   **Figure 5** shows coastal management at Lyme Regis in Dorset.

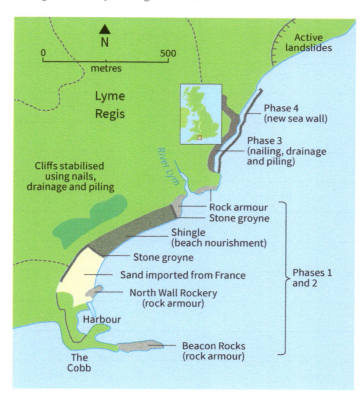

Figure 5

Assess the relative costs and benefits of managing coasts like in **Figure 5** using hard-engineering management strategies.

(8)

_____

_____

_____

_____

_____

_____

_____

_____

_____

_____

_____

_____

_____

_____

_____

_____

_____

_____

_____

_____

_____

_____

_____

(Total for Question 4 = 8 marks)
TOTAL FOR SECTION A = 27 MARKS

Answer ALL questions. Write your answers in the spaces provided.

5 (a) **Figure 6** shows the population structure of the Outer Hebrides in 2004 and in 2020, an extreme, rural part of the United Kingdom.

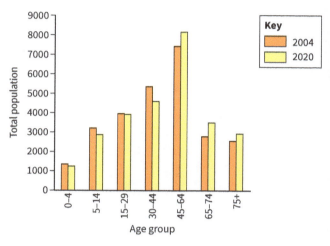

Figure 6

(i) Identify the modal age group in both 2004 and 2020.

(1)

_____

(ii) Suggest **one** way in which the population structure of the Outer Hebrides could affect economic activity there.

(2)

_____

_____

_____

_____

(iii) Explain **one** attempt to improve the economies of rural areas.

(2)

_____

_____

_____

_____

(Total for Question 5 = 5 marks)

Dynamic UK Cities

6 (a)   **Figure 7** shows part of a city in the UK.

Figure 7

(i)   Identify the part of the city shown in **Figure 7**.

(1)

☒ **A**   CBD
☒ **B**   Inner city
☒ **C**   Suburbs
☒ **D**   Rural–urban fringe

(ii) Describe **two** features that makes this part of the city distinctive.

(2)

_____

_____

_____

_____

(iii) Explain **two** reasons why areas such as that shown in **Figure 7** may need to be re-branded.

(4)

1 _____

_____

_____

_____

2 _____

_____

_____

_____

(b)   **Figure 8** shows a map of the park and ride schemes for the city of Bath in south-west England.

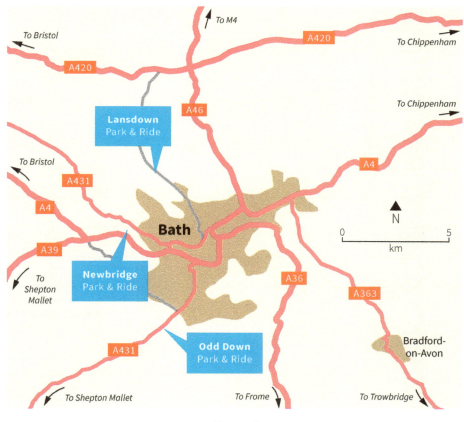

Figure 8

(i) Suggest **one** way in which park and ride schemes may affect cities such as Bath.

(3)

_____

_____

_____

_____

_____

_____

_____

_____

(ii) Apart from park and ride schemes, explain **two** ways in which people attempt to make urban living more sustainable.

(4)

1 _____

_____

_____

_____

_____

2 _____

_____

_____

_____

_____

(Total for Question 6 = 14 marks)

Investigating a UK Geographical Issue

*Spelling, punctuation, grammar and the use of specialist terminology will be assessed in this question.

*7 **Figure 9** compares wealth and deprivation in two different parts of the city of Birmingham.

	Sparkbrook	Sutton Four Oaks	*Birmingham average*
Unemployment (%)	24.5	3.1	*12.0*
Economically active or at work (%)	48	81	*68*
Working age population with no qualifications (%)	49.7	20.9	*37.1*
Pupils with five GCSEs, A*–C (%)	51	74	*58*
Children living in poverty (%)	49	7	*34*
Average household income (£)	21 000	40 000	*31 000*
Households with income less than £15 000 (%)	46	12	*27*
Households with income over £35 000 (%)	12	47	*27*

Figure 9

In this question, up to 4 additional marks will be awarded for your spelling, punctuation and grammar and your use of specialist terminology.

Assess the extent to which there is inequality between different parts of major UK cities, such as Birmingham.

(8 and 4)

_____
_____
_____
_____
_____
_____
_____
_____
_____
_____
_____
_____

(Total for Question 7 = 12 marks)
TOTAL FOR SECTION B = 31 MARKS

Section C1 Geographical Investigations: Fieldwork in a Physical Environment

Answer EITHER Question 8 OR Question 9 in this section.
Write your answers in the spaces provided.

If you answer Question 8 put a cross in this box ☐.

Investigating Coastal Change and Conflict

8   You have carried out a fieldwork investigation in a coastal environment.

Name your coastal environment location:

_____

Figure 10 shows some fieldwork equipment that could be used when carrying out a coastal investigation.

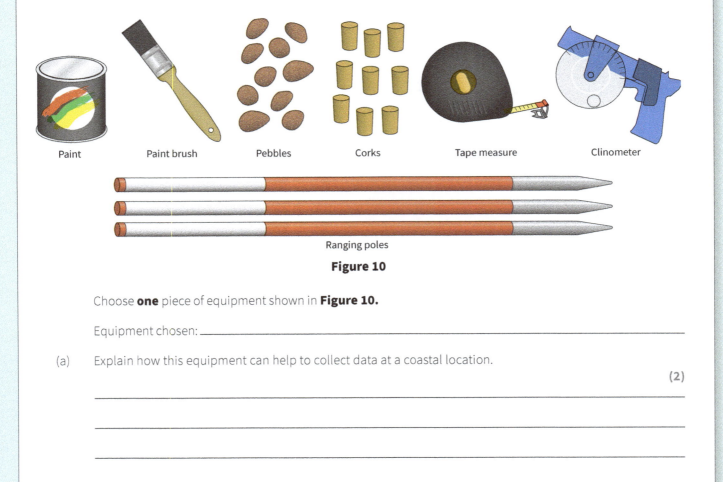

| Paint | Paint brush | Pebbles | Corks | Tape measure | Clinometer |

Ranging poles

Figure 10

Choose **one** piece of equipment shown in **Figure 10**.

Equipment chosen: _____

(a)   Explain how this equipment can help to collect data at a coastal location.

(2)

_____

_____

_____

_____

(b)   **Figure 11** shows a graph of pebble sizes from four samples taken at five locations along a beach.

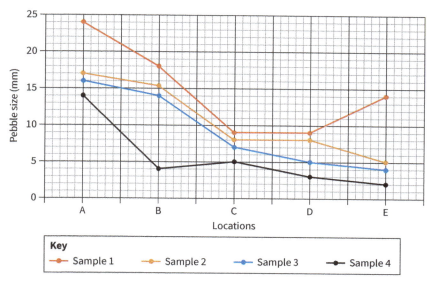

Figure 11

(i)   State **one** reason why this is an **inappropriate** graph for this type of data.

(1)

_____

_____

(ii) State **one** more suitable form of graph to illustrate this type of data.

(1)

_____

_____

(c)   (i)   Explain **two** challenges that the students might face in trying to make their sediment data collection as accurate as possible.

(4)

1 _____

_____

_____

_____

2 _____

_____

_____

_____

(ii) Explain **one** concept or theory that students collecting data might be trying to test.

(2)

_____

_____

_____

_____

(iii) Assess the accuracy and reliability of your conclusions in your coastal fieldwork.

(8)

_____

_____

_____

_____

_____

_____

_____

_____

_____

_____

_____

_____

_____

_____

_____

_____

(Total for Question 8 = 18 marks)

If you answer Question 9 put a cross in this box ☐.

Investigating River Processes and Pressures

9 You have carried out a fieldwork investigation in a river environment.

Name your river environment location:

_____

Figure 12 shows some fieldwork equipment that could be used when carrying out a river investigation.

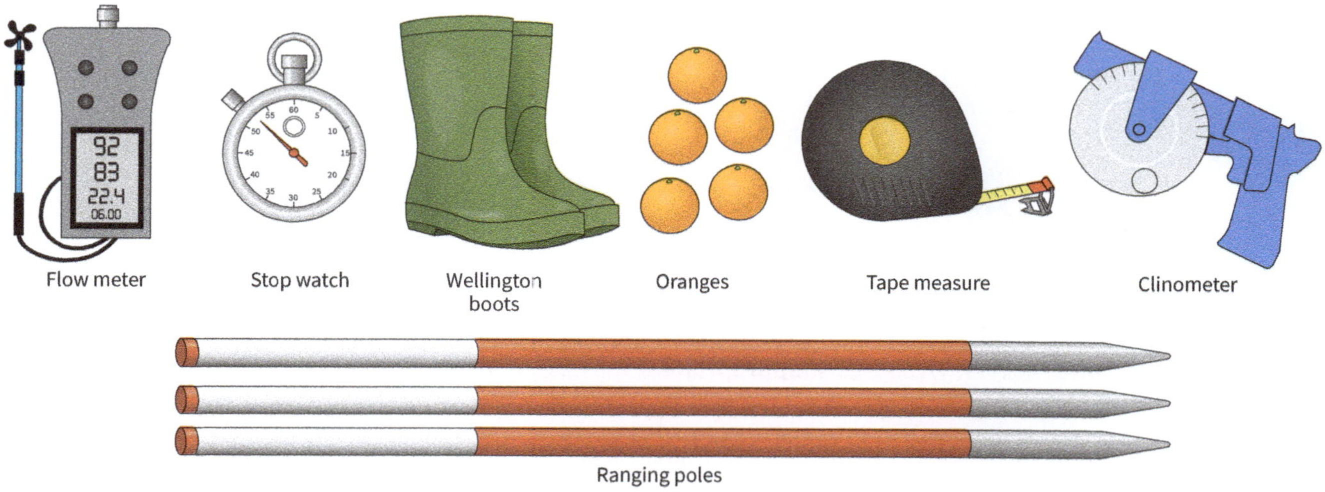

| Flow meter | Stop watch | Wellington boots | Oranges | Tape measure | Clinometer |

Ranging poles

Figure 12

(a) Choose **one** of the pieces of equipment shown in **Figure 12.**

Equipment chosen: _____

Explain how this equipment can help to collect data at a location along a river.

(2)

_____

_____

_____

_____

(b)   **Figure 13** shows a graph of velocity of a river by taking measurements four times at five different locations.

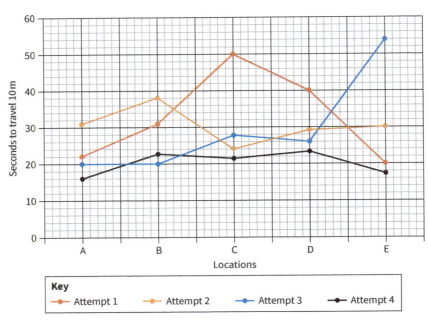

Figure 13

(i)   State **one** reason why this is an **inappropriate** graph for this type of data.

(1)

_____

_____

(ii) State **one** more suitable form of graph to illustrate this type of data.

(1)

_____

_____

(c)   (i)   Explain **two** challenges that students might face in trying to make their river data collection as accurate as possible.

(4)

1 _____

_____

_____

_____

2 _____

_____

_____

_____

(ii) Explain **one** concept or theory that students collecting data might be trying to test.

(2)

_____

_____

_____

_____

(iii) Assess the accuracy and reliability of your conclusions in your river investigation.

(8)

_____

_____

_____

_____

_____

_____

_____

_____

_____

_____

_____

_____

_____

_____

_____

_____

_____

(Total for Question 9 = 18 marks)
TOTAL FOR SECTION C1 = 18 MARKS

Section C2 Geographical Investigations: Fieldwork in a Human Environment

**Answer EITHER Question 10 OR Question 11 in this section.
Write your answers in the spaces provided.**

If you answer Question 10 put a cross in this box ☐.

Investigating Dynamic Urban Areas

10 (a)   A group of students investigated people's views on the quality of life in an urban area.

They collected primary data using a questionnaire, choosing a method where they questioned every 10th person that walked past them.

(i)   Identify which **one** method of sampling the students used.

(1)

☒ **A** Random
☒ **B** Stratified
☒ **C** Systematic
☒ **D** Opportunistic

(ii) The first two questions on the questionnaire required the students to note the age and sex of the person they were questioning.

Explain **one** reason why this information might prove helpful in increasing the accuracy of their findings.

(2)

_____

_____

_____

_____

(b)   **Figure 14** shows some secondary information about unemployment, which they
        included in their write up of their geographical investigation.

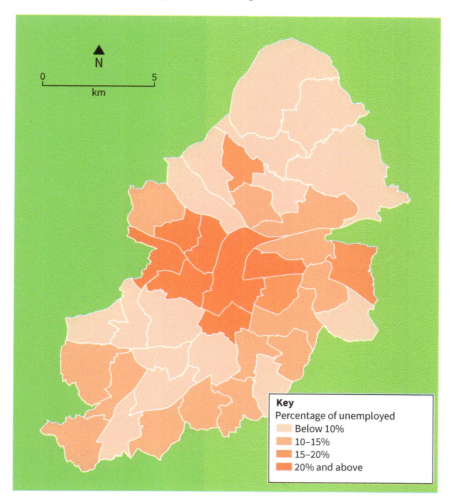

Figure 14

(i)   Identify the method of data presentation technique used in **Figure 14**.

(1)

_____

_____

(ii)   Explain **one** weakness of this technique when showing unemployment in an urban area.

(2)

_____

_____

_____

_____

(c)   **Figure 15** is a bi-polar diagram that a student constructed using both primary and secondary data from two different parts of an urban area.

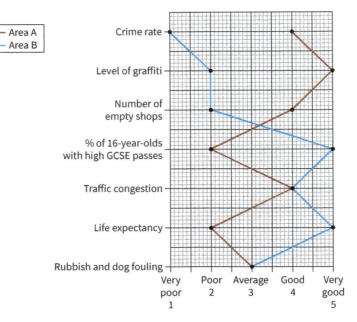

Figure 15

Explain **two** pieces of evidence in **Figure 15** to show that there are social inequalities in this urban area.

(4)

1 _____

_____

_____

_____

2 _____

_____

_____

_____

(d)   Assess the usefulness of your data collection methods in helping you to investigate variations in urban quality of life.

(8)

_____

_____

_____

_____

_____

_____

_____

_____

_____

_____

_____

_____

_____

_____

_____

_____

_____

_____

(Total for Question 10 = 18 marks)

If you answer Question 11 put a cross in this box ☐.

Investigating Changing Rural Areas

11 A group of students investigated people's views on the quality of life in a rural area.

They collected primary data using a questionnaire, choosing a method where they questioned every 10th person that walked past them.

(a) (i) Identify which **one** method of sampling the students used.

(1)

⊠ **A** Random
⊠ **B** Stratified
⊠ **C** Systematic
⊠ **D** Opportunistic

(ii) The first two questions on the questionnaire required the students to note the age and sex of the person they were questioning.

Explain **one** reason why this information might prove helpful in increasing the accuracy of their findings.

(2)

_____

_____

_____

_____

(b)   **Figure 16** shows some secondary information about the provision of services in some Dorset villages, which they included in their write up of their geographical investigation.

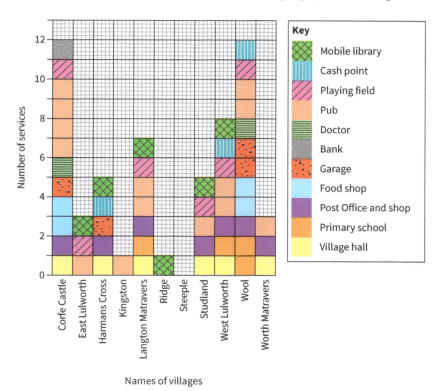

Figure 16

(i)   Identify the data presentation technique used in **Figure 16**.

(1)

_____

_____

(ii)   Explain **one** weakness of this technique when showing provision of services in a rural area.

(2)

_____

_____

_____

_____

(c) **Figure 17** is bi-polar diagram that a student constructed using data from two different parts of a village in Essex.

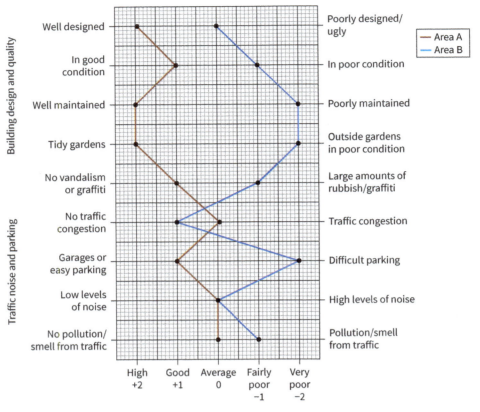

Figure 17

Explain **two** pieces of evidence in **Figure 17** that shows that there are social inequalities in this village.

(4)

1 _____

_____

_____

_____

2 _____

_____

_____

_____

_____

(d) Assess the usefulness of your data collection methods in helping you to investigate variations in rural quality of life.

(8)

_____

_____

_____

_____

_____

_____

_____

_____

_____

_____

_____

_____

_____

_____

_____

_____

_____

_____

_____

(Total for Question 11 = 18 marks)
TOTAL FOR SECTION C2 = 18 MARKS
TOTAL FOR PAPER = 94 MARKS

GCSE 9-1 Geography Edexcel B
Practice Paper 3

People and Environment Issues –
Making Geographical Decisions

Time allowed: 1 hour 30 minutes
Total number of marks: 64 (including 4 marks for spelling,
punctuation, grammar and use of specialist terminology (SPaG))

Instructions
Answer **all** questions.

Section A People and the Biosphere

Answer ALL questions. Write your answers in the spaces provided

Some questions must be answered with a cross in a box ☒. If you change your mind about an answer, put a line through the box ☒ and then mark you new answer with a cross

1   Use Section A in the Resource Booklet to answer these questions.

(a)   (i)   Identify which **one** of the following is **not** a factor that can alter a global biome **locally**.

(1)

☒ **A**   Latitude
☒ **B**   Soil type
☒ **C**   Drainage
☒ **D**   Rock type

(ii)   **Figure 1** shows an example of a small-scale ecosystem.
Name **one** abiotic and **one** biotic component shown on **Figure 1**.

(2)

Abiotic _____   Biotic _____

(iii)   Explain how **one** local factor has altered the characteristics of the global biome at this location.

(2)

_____

_____

_____

_____

(b)   **Figures 2** and **3** show global population growth in 2015 and the global distribution of biomes.

(i)   Identify which continent shows the greatest rate of population growth.

(1)

_____

(ii)   Identify the **two** biomes likely to be most affected by this high growth rate.

(2)

1 _____

2 _____

(Total for Question 1 = 8 marks)
TOTAL FOR SECTION A = 8 MARKS

Section B Forests Under Threat

2   Use **Section B** in the Resource Booklet to answer these questions.

(a)   **Figure 4** shows a vegetation map of Democratic Republic of the Congo (DRC), a country in Africa.

(i)   Define the term 'primary forest'.

(1)

_____

_____

(ii)   Describe the distribution of primary forest within DRC.

(2)

_____

_____

_____

_____

(b)   Study **Figure 5**. Suggest why the loss of forest cover in DRC is important
i) for DRC, and ii) globally.

(4)

Importance for DRC: _____

_____

_____

_____

Importance globally: _____

_____

_____

_____

(c)   Study **Figure 6**.
(i)   identify the population of DRC in 2000.

(1)

Answer = _____

(ii)   Identify the year in which the population of DRC is likely to reach 200 million.

(1)

Answer = _____

(Total for Question 2 = 9 marks)
TOTAL FOR SECTION B = 9 MARKS

Section C Consuming Energy Resources

3 Use Section C in the Resource Booklet to answer these questions.

(a) Study **Figure 7**, which gives information about the lifestyle of indigenous peoples of the rainforest in DRC.
Explain **two** ways in which the lifestyle of indigenous peoples can be considered sustainable.

(4)

1 _____

_____

_____

_____

2 _____

_____

_____

_____

(b) Study **Figure 8**. 98% of population of DRC use firewood as their source of energy in their homes.
Explain why, **apart from deforestation**, the following two factors are important because of the over-reliance on the use of firewood.

(4)

Impact on people's health: _____

_____

_____

_____

Impact on woman's place in society: _____

_____

_____

_____

(c)   (i)   Explain **one** reason why, in a developing country like DRC, demand for energy
              is almost certain to increase.

                                                                                            (3)

_____

_____

_____

_____

_____

_____

     (ii)   **Figure 9** shows the cost and benefits of developing HEP **as a source of energy**
              in DRC.
              Assess the suitability of developing HEP in DRC.

                                                                                            (8)

_____

_____

_____

_____

_____

_____

_____

_____

_____

_____

_____

_____

_____

_____

_____

     (iii) Study **Figure 10**, a photograph of Inga 1 and 2 HEP installations on the Congo River.
              Suggest **two** impacts of constructing HEP installations on the rainforest environment.

                                                                                            (4)

1 _____

_____

_____

_____

2 _____

_____

_____

_____

(iv) **Figure 11** gives information about the proposed Inga 3 HEP project and four different views about it. Assess the reasons why there is such a range of views about building this project.

(8)

_____

_____

_____

_____

_____

_____

_____

_____

_____

_____

_____

_____

_____

_____

_____

_____

_____

(Total for Question 3 = 31 marks)
TOTAL FOR SECTION C = 31 MARKS

Section D Making a Geographical Decision

*In this question, 4 of the marks awarded will be for your spelling, punctuation and grammar and your use of specialist terminology.

*4 Study the **three** options below for how DRC should provide sufficient energy to provide for its population and to further develop the country in the 21st century.

> **Option 1**: Cease building large-scale, top-down, energy generating schemes and encourage more small-scale, bottom-up projects.

> **Option 2**: Adopt more sustainable management of DRC's rainforests including replanting to replace deforestation caused by building energy installations.

> **Option 3**: Expand large-scale HEP generation as the only practical way of solving DRC's energy needs.

Select the option that would be the best plan to ensure **long-term** energy security for the economy, and the environment.

Use information from the Resource Booklet, and knowledge and understanding from the rest of your geography course, to support your answer.

(12 and 4)

Chosen option: _____

_____

_____

_____

_____

_____

_____

_____

_____

_____

_____

_____

_____

_____

_____

_____

_____

_____

_____

_____

_____

_____

_____

_____

_____

_____

_____

_____

_____

_____

_____

_____

_____

_____

_____

_____

_____

_____

_____

(Total for Question 4 = 16 marks)
TOTAL FOR SECTION D = 16 MARKS
TOTAL FOR PAPER = 64 MARKS

Figure 1

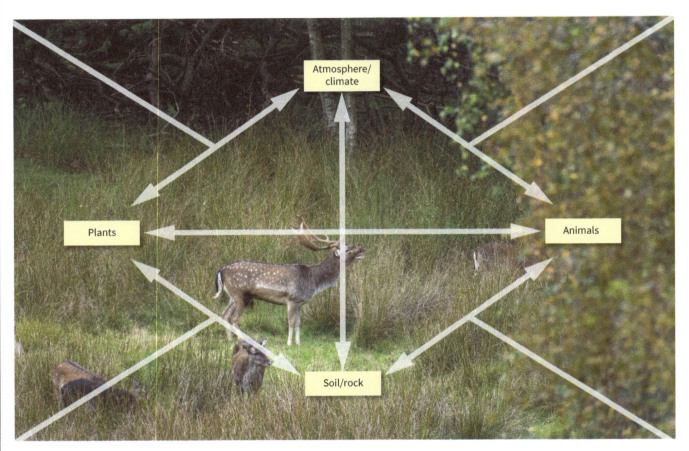

A small-scale ecosystem

Figure 2

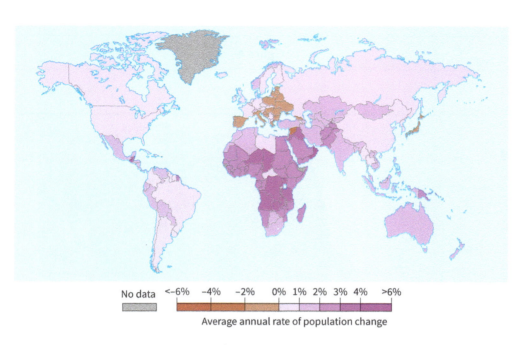

No data | <-6% −4% −2% 0% 1% 2% 3% 4% >6%

Average annual rate of population change

Population growth rate, 2020

Figure 3

Key

- Tundra
- Coniferous forest
- Temperate deciduous forest
- Temperate grassland
- Mediterranean
- Desert
- Tropical rainforest
- Tropical grassland (savanna)
- Other biomes (e.g. polar, ice, mountains)

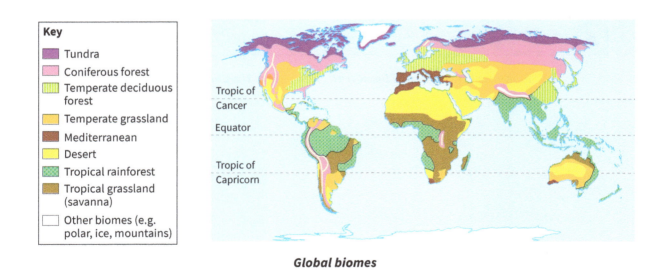

Tropic of Cancer

Equator

Tropic of Capricorn

Global biomes

Figure 4

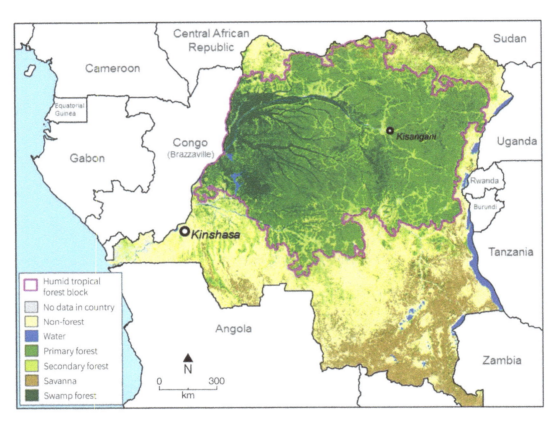

Rainforest cover map of Democratic Republic of the Congo (DRC)

Figure 5

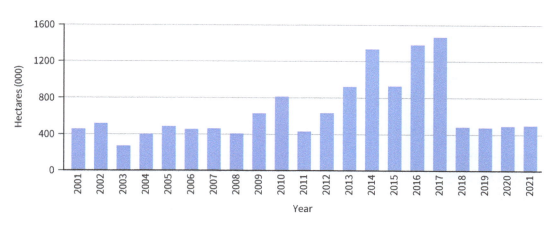

Tree cover loss in DRC

Figure 6

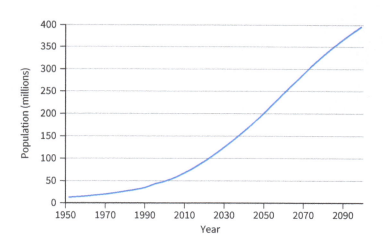

Population growth in DRC, 1950–2100

DRC has one of the highest population growth rates in the world

---

Figure 7

Small populations of indigenous people living in rainforest in DRC rely on it for their livelihood.

Patches of rainforest are cleared to plant cassava, corn and yams for themselves and peanuts, rice, coffee beans and oil palms to sell.

They are hunter-gatherers, hunting animals and gathering foods from the rainforest.

Meat is exchanged between farmers for vegetables, grains and other produce.

They collect caterpillars, grubs and honey to eat along with a variety of plants that they use for food and medicine.

Shelters are made from young trees woven together.

Lifestyle of indigenous people of the rainforest

Figure 8

Firewood is an important source of energy in DRC

Figure 9

Makes use of the ample supply of water in DRC's dense river system.

Will flood large areas of rainforest destroying ecosystems.

There are plenty of sparsely populated sites for the construction of the dams.

Dams become blocked with soils washed down from deforested slopes.

The dams often have a relatively short life because of lack of maintenance.

HEP: Good or Bad

Submerged forests rot making the water acidic. This attacks the turbines meaning frequent replacement.

Provides a regular supply of electricity to supply local and national needs, with the potential to export.

Could be multi-purpose.

Very expensive to build, and is likely to be dependent on top-down IGO development aid.

Relatively cheap to run after the construction phase and is renewable.

The country has a high rainfall and relatively short dry season.

Reduces the over-reliance on the use of firewood as its main source of energy.

Costs and benefits of developing HEP as an energy source for DRC

Figure 10

Inga 1 and 2 dams on the Congo River

Figure 11

Construction of world's largest dam in DRC could begin within months

DRC is experiencing an energy crisis because of a lack investment and management in the energy sector. 93.6% of the country is dependent on wood fuel as its main source of energy, resulting in deforestation and environmental degradation. The solution is held to be Inga 3, the largest HEP dam in the world. The project, costing about US$100 billion, could eventually span the Congo River, the world's second largest river by volume. When built, it is expected to have the capability to generate nearly twice as much electricity as the Three Gorges Dam in China. It is being jointly financed by the World Bank, the DRC government and the South African government.

View 1
The World Bank Group is committed to supporting DRC in providing affordable and reliable energy for its people and to drive sustainable growth for the economy. The Bank will remain involved by focusing on rehabilitating mid-size HEP plants, increasing electricity access to the population, and supporting regional electricity transmission.
World Bank, 2016

View 2
In March 2022, 26 traders and shoppers were electrocuted at the Matadi-Kibala food market in Kinshasa, DRC, when a high-voltage power cable snapped and its live end landed in a water-filled ditch. Those who were killed, mostly impoverished women traders, were unlikely to have electricity at home. 'People see the power lines above them, but they don't have access to that electricity' said Kambale Musavuli, a human rights activist and analyst at the Center for Research on the Congo-Kinshasa, which is based in Kinshasa and the United States.
New Frame, 2022

View 3
Governments of DRC and South Africa signed a treaty to develop Inga 3, one of Africa's largest dam projects. South Africa will be the main buyer of power generated at the dam. But there has been little progress. Meanwhile, the majority of people in DRC continue to lack access to energy – estimates remain low at between 9 and 15% of the population.
Rudo Sanyanga (2016). She is from Zimbabwe, and has a PhD in Aquatic Systems Ecology from Stockholm University and is the Africa Program Director of International Rivers.

View 4
The two parts of Inga 3 are expected to displace about 35 000 people (when land behind the dam is flooded) as well as greatly affecting the river's fish supplies. Campaigners argue that the power generated by the HEP project will travel so far to reach DRC's industrial mines, and then a further 5000 km all the way to South Africa, that the majority of DRC's population not served by the nation's gridlines will be bypassed.
International Rivers, a California-based NGO (2017)

Answer guidance for On your marks! questions

1   Mopping up the 1-mark questions

Page 15 Now try these!

1   B
2   D
3   D

Pages 16–17 Now try these!

1(a) Graph should show an incline from 0 to 900 for tropical rainforest and again from 900 to 1800 for temperate forest

Both should be done for 1 mark – no half marks if only one completed correctly

1(b) Coniferous forest exists between 1800 metres and 3800 metres above sea level

1 mark for both correct – no half marks

Allow a range of 3600–3900 for the second figure

1(c) Allow a range between 4600 and 4900 metres. Answer does not need to say 'above sea level'

2(a) B

2(b) North West. No other answer is acceptable

2(c) 1 mark for any of the following:
- reference to 'quarry'
- reference to 'tips'
- reference to 'mine'
- reference to 'shafts'
- any other acceptable piece of evidence

Pages 18–19 Now try these!

1   1 mark for any of the following:
- formed in layers / layered
- formed in river deposits / sea or coastal deposits / glacial deposits / wind-blown deposits
- any lithology characteristics, e.g. 'it has joints / bedding planes'
- any other correct characteristic

2   1 mark for any of the following:
- freeze-thaw (accept 'frost shattering')
- breakdown by plant roots / allow 'biological weathering'

3   1 mark for any of the following:
- allow any general land uses e.g. industry, housing, forestry

- specific activities e.g. mining, factories, hotels, office blocks
- do not allow types of activities that are not linked to the landscape e.g. fishing

4   1 mark for any of the following:
- change in the Earth's orbit / orbital changes / Milankovitch cycles
- variations in solar output / heat output from the sun / sunspots
- volcanic activity
- asteroid collisions

5   1 mark for any of the following:
- ice cores
- tree rings
- historical evidence such as paintings
- historical evidence such as diaries

6   1 mark for a definition similar to the following:
- 'where more people are leaving a city to live in smaller towns or settlements than are entering it'
- 'population movement has more people leaving the city to live elsewhere than arriving'
- no marks for just 'people leaving'

7   Human Development Index (all three words must be correct – no half marks)

8   1 mark for any of the following:
- a bend in the river
- slip-off slope
- river cliff
- erosion on the outer bend
- any other acceptable feature

9   1 mark for any of the following:
- people may be killed / injured
- homes may be damaged / destroyed
- loss of power supplies / water
- any other answer which implies direct damage caused by the tectonic event itself

10   1 mark for any of the following:
- a resource which is infinite
- a resource that comes from natural sources such as wind or sun
- a resource that will never run out

11 Erosion is the wearing away and removal of material / rocks by a river / glacier / wind
- allow only 'wearing away of rocks', but not just 'wearing away'

2 Maxing out the 2-mark questions

Page 21 Now try these!

1 1 mark for the feature named and 1 mark for a developed point. Example:
- 'There may be housing shortages / housing may be overcrowded and of poor quality' (1) 'meaning that health may be poor' (1)
- 'Waste disposal services may be poor' (1) 'meaning that disease spreads more easily and health is worse' (1)
- 'Inequality may be high' (1) 'because some people are very wealthy and have a high quality of life' (1)
- any other acceptable feature of quality of life that is developed

2 1 mark for the feature named and 1 mark for a developed point. Example:
- 'The centre of the storm is calm' (1) 'with calm conditions and clear skies' (1)
- 'Fastest winds circulate around the storm' (1) 'with very low pressure air' (1)

3 1 mark for the measure named and 1 mark for explanation of how it measures development No marks for a social or political indicator. Allow HDI as it includes GNI. Example:
- 'GDP / GNI' (1) 'because it measures the wealth produced by a country in a year' (1)

4 1 mark for benefit given and 1 mark for explanation of how it benefits the country. Example:
- 'There could be investment in industries / factories' (1) 'which increase the number of jobs for people' (1)
- 'There could be investment in services' (1) 'which increases the number of high-paid jobs for people' (1)

Page 22 Activity 1

Answer 1 1 mark; the point is not developed so no second mark

Answer 2 2 marks for rising sea level (1) with a large range of predictions (1)

Answer 3 0 marks as 'changing' is not specific enough – it is not the same as 'rising'; 'ice caps are melting' is an explanation not a description of future sea level rise

Answer 4 2 marks for increases have a wide range (1) and use of data to illustrate (1)

Page 23 Now try these!

1 1 mark for the feature and 1 mark for a developed descriptive point. Example:
- 'The cloud is circulating' (1) 'in an anti-clockwise direction' (1)
- allow 'thick cloud' but not 'a lot of rain'

2 1 mark for a difference and 1 mark for developed point (e.g. by including data). Example:
- 'More people live further away from work in a megacity' (1) 'e.g. only half are within 15 minutes of work compared to 82% in the rest of the country' (1)

Page 25 Now try these!

1 1 mark for the way / method of preparing for the hazard and 1 mark for a developed point. Example:
- 'They can set up seismometers' (1) 'so that earth tremors before an eruption can be detected' (1)
- 'They can develop evacuation plans' (1) 'so that people know where to go if there's an eruption' (1)

2 1 mark for the transport feature described in a named UK city and 1 mark for a developed point. Example:
- 'London has introduced congestion charging' (1) 'so that people find it cheaper to use public transport' (1)
- 'London has introduced electric and hydrogen buses' (1) 'so that CO_2 emissions are reduced' (1)

3 1 mark for the feature and 1 mark for a developed point. Example:
- 'Plates pull apart' (1) 'so that magma rises to the surface as lava' (1)
- 'Lava erupts out of the crack / space between the two plates' (1) 'so that volcanoes form there' (1)

Answer guidance

4 1 mark for the cause and 1 mark for developed explanation. Example:
- 'Houses / infrastructure may not be so well built' (1) 'so that strong winds cause them to collapse' (1)
- 'Warning systems may not be so well-developed' (1) 'meaning that fewer people took precautions to protect their homes' (1)

Pages 26–27 Now try these!

1 1 mark for the method and 1 mark for a developed point. Example:
- 'Setting up weather satellites' (1) 'which are used to track a tropical cyclone / forecast its arrival' (1)
- 'Building raised shelters' (1) 'which are buildings used to house evacuated people' (1)

2 1 mark for the way / method and 1 mark for the detailed explanation. Example:
- 'Setting up weather satellites' (1) 'so that people can predict when a tropical cyclone might arrive' (1)
- 'Building shelters' (1) 'to give somewhere safe to go during the cyclone' (1)

3 1 mark for the adaptation named and 1 mark for a developed point. Example:
- 'Trees have buttress roots' (1) 'which spread outwards from the tree' (1)
- 'Leaves have pointed tips' (1) 'so that water can run off easily' (1)

4 1 mark for the adaptation named and 1 mark for the detailed explanation. Example:
- 'Trees have buttress roots' (1) 'which support the weight of the tree' (1)
- 'Tall emergent tress grow high above the rest' (1) 'which allows them to compete for light' (1)

3 Tackling the 3-mark questions

Page 29 Now try these!

1 1 mark for a feature and 2 marks for a developed explanation of the difference. Example:
- 'Volcanoes at divergent boundaries are gently sloping' (1) 'whereas those at convergent boundaries are steeply sloping (1) 'because of the difference in lava temperatures' (1)

2 1 mark for each of a sequence of changes brought by a change of land use. Example:
- 'Changing from farmland to urban land removes permeable soil' (1) 'and replaces it with impermeable concrete or tarmac' (1) 'which means that surface runoff is much faster' (1)

3 1 mark for an impact and 2 marks for a developed explanation of the difference. Example:
- 'Economic development leads to an increase in urbanisation' (1) 'because there is more employment in urban areas' (1) 'as industrialisation leads to greater numbers of factories' (1)

4 Managing the statistics questions

Pages 30–31 Now try these!

1(a) 28.3 degrees. Units are not needed to award the mark

1(b) 12 degrees. Units are not needed to award the mark

1(c) 345.4 mm. Units are not needed to award the mark but answer must be exact

1(d) 87.6% (1) Working should show amount fallen in rainy season June–Sept as 302.5 (1) divided by the total and multiplied by 100 (1)

2(a) 6

2(b) 166 billion (must say 'billion' to gain the mark)

Page 33 Now try these!

1(a) Median Y; Lower quartile Z; Upper quartile X

1(b) 68 (1) or 49 (1)

5 Excelling at 4-mark questions

Page 35 1 Plan your answer

(a) Any two from the following:
- buildings / houses have collapsed
- roofs have been removed
- weak building materials
- any other acceptable response

(b) Any two from the following:
- low GDP means poor building materials may have been used
- lack of government regulations about building quality

- strength of winds in a tropical cyclone
- any other acceptable response

Page 36 2 Mark an answer
Refer to the mark scheme given on page 36

Page 37 3 Mark a different answer
This answer gets 2 marks: 'wooden shacks like the one in the photo would not be able to stand up to strong hurricane winds' (1) and 'so they would fall down' (1). The second part of the answer however is not relevant

Pages 38–39 1 Plan your answer
Evidence: two ways in which HDI data helps to understand a country's level of development

Box and explain the command word: 'explain' means gives detail about how it helps to understand development

Focus: the question is about HDI, not any other measure of development

What you have to write: two measures of development included in HDI and what they tell you about the country

(a) Any two from the following:
- GDP / gross domestic product
- life expectancy
- mean years of schooling / literacy rate

(b) For GDP: tells you about the wealth / income of a country as it shows how much a country earns on goods and services

For life expectancy: tells you about the healthcare / how much is spent on healthcare / how healthy a population is as a whole

For mean years of schooling: tells you about the education system / how much a country spends on schools / how well educated the population is

Page 39 2 Write your answer
1 'HDI shows a country's level of development because it shows GNI / GDP' (1) 'which tells you how wealthy a country is' (1)
2 'HDI shows a country's level of development because it includes life expectancy (1) which shows the quality of healthcare within a country' (1)

Page 40 3 Mark your answer
Refer to the mark scheme given on page 40

Page 41 4 Mark a different answer
This answer gets 4 marks because the candidate has:
- stated two things that HDI includes
- explained what each element of HDI shows

Page 41 5 Improve an answer
- Briefly define / explain what life expectancy measures (1) with its relationship to health care (1)
- Explain a second measure included in HDI and then explain what this tells you about development

Page 42 Now try this!
Answer should include:
- two uses of Figure 1 – e.g. the relationship between headlands (including their names) and areas of hard rock type (limestone / chalk); the relationship between bays (including their names) and areas of softer rock types (clays and sands)
- two examples of your understanding – e.g. explaining how / why softer rocks would be eroded more than harder rocks

Page 44 1 Plan your answer
(a) Make sure that the two text boxes each contain e.g.:
- a method of hard engineering (examples could include sea wall, revetment, gabion, rock armour / rip-rap, or stone / wooden groyne
- an explanation about how it protects the coastline

Page 44 2 Mark an answer
This answer gets 3 marks. Part 1 names groynes (1) and shows a clear understanding of their purpose in preventing erosion (1). Part 2 gets a mark for sea wall (1) but the explanation does not explain how it stops erosion

Page 45 4 Mark your answer
(a) Look for two reasons which show a clear understanding of how urban land use can lead to a high risk of flooding

An example answer could be: Urban land uses change the surface from permeable soil to impermeable concrete (1) so there is more surface runoff (1). Urban areas have storm drains (1) which move water quickly into streams and rivers (1)

(b) Award an appropriate mark

Pages 46–49 Now try these!

1   Answer should:
- include reference to both derelict land and Glasgow's most deprived areas
- suggest a reason for derelict land, e.g. closed industries / deindustrialisation
- suggest a reason for the deprivation found in the same areas of the city, e.g. loss of jobs when industries closed
- refer to a least one piece of evidence on the map (Figure 1)

2   Answer should include:
- two impacts of urban growth, e.g. increased rural–urban migration, or rapid increase in the urban population
- an explanation of how this has affected quality of life, e.g. insufficient housing or housing shortages (1), or insufficient infrastructure / water supply / electricity supply (1), or high rents / costs of housing (1), or cramped / overcrowded accommodation (1)
- reference to at least one feature of the photograph in Figure 2

3   Answer should include:
- reference to Figure 3 which shows a reduction in Arctic sea ice by 50% (1) compared to the 1979–2000 average (1)
- an explanation that this reduction in sea ice is a result of climate change (1)
- an explanation of how climate change is reducing sea ice, including the link to rising global temperatures (1)

6   Getting your head around the 8- and 12-mark questions

Page 52 1 Plan your answer

Refer back to the guidance on page 35 for how to BUG the question. Your four boxes could include:

Evidence: the problems and benefits of living in a low-income area of a megacity in developing or emerging countries – ones that you know about and ones that you can see from the photo; from the photo you might pick out the poor quality housing

Box and explain the command word: 'Assess' means that you need to judge whether or not the statement in the question is true or not and why

Focus: problems and benefits of living in low-income areas of a megacity in a developing or emerging country

What you have to write: three points overall:
- one way in which the statement might be true – supported with evidence (e.g. the problems of housing seen in the photo)
- one way in which the statement might not be true – supported with evidence (e.g. the benefits of living in a megacity, such as jobs)
- a third paragraph that discusses either another problem or another benefit – depending on what judgement you will make about the statement
- decide in a sentence whether you agree with the statement in the question or not – this acts as your mini-conclusion

Page 54 3 Write your answer

Answer should include:
- three points overall:
  - one problem of living in a low-income area in a megacity
  - one benefit of living in a low-income area in a megacity
  - another paragraph about either another problem OR another benefit of living in a low-income area in a megacity
- a judgement about whether or not the problems are worse than the benefits – are you agreeing or disagreeing with the statement in the question?

Page 56 5 Mark different answers

Sample answer 1 The candidate uses a good PEEL structure with their judgement sentences at the end of each paragraph ensuring that they link back to the question. For example:

Point: agrees with the statement and talks about Mumbai as a megacity that has grown quickly; states the problem that the rapid growth has meant that services like water can't keep up

Evidence: uses the evidence of the building materials seen in Figure 1

Explanation: explains that they agree with the statement because there are many problems and so anyone with more income would not choose to life in low-income areas such as Dharavi

They do this again for their second point – ensuring that their answer is well developed:

Point: mentions the informal settlements of Mumbai and how people build their own shelters illegally

Evidence: how these informal settlements are built on sloping land unsuited to building

Explanation: explains how this makes people vulnerable and how landslides can result; they explain how they agree with the statement because people with higher-incomes would not live on land like this

The candidate gives a clear view of what living in a low-income area of a megacity might be like and expands their points to give detail. They make a judgement and so gain 7 marks, a Level 3 answer. The answer is not perfect as it is a little short on data

Sample answer 2 This answer gets a mark in the middle of Level 2 (5 marks) because although they give a point and evidence, their explanation is weaker. For example:

Point: recognises that some may agree with the statement because people living in informal settlements have lots of problems

Evidence: refers to the dangers of electricity from bare wires

Point: agrees with the statement that cities provide jobs for many people

Evidence: people move from the countryside and get better jobs than they would have in rural areas

Point: states that cities like Mumbai have lots or schools and maybe universities too

Evidence: there are also hospitals and medical treatment centres too

However, the candidate does not give a strong judgement and does not explain their points enough. For example, in their last point, they don't explain why this is a benefit. They give a single statement at the end of their answer about living in Mumbai but an agreement about whether they agree with the statement or not is not built up

throughout the answer. They also don't include enough specific explanation or evidence about what they know – they mention general statements about living in Mumbai without much detail

Page 59 1 Plan your answer

Refer back to the guidance on page 35 for how to BUG the question. Your four boxes could include:

Evidence: three impacts of international migration – at least one from the map in Figure 1 and at least one from the major UK city you have studied

Box and explain the command word: 'Assess' means that you need to judge whether or not international migration has had an impact on UK cities, e.g. positive or negative

Focus: the impacts of international migration on major UK cities – including London (shown in the map)

What you have to write: impacts of international migration – how have migrants from abroad affected cities in the UK?

Page 61 3 Write your answer

Answer should include:

- three impacts of international migration for UK cities
- at least one impact linked in evidence from Figure 1 (the map of London)
- at least one impact from a city that you have named in your answer
- a judgement about the impact of international migration on major UK cities

Page 63 5 Mark different answers

Sample answer 1 The candidate uses a PEEL structure but the link (and so their judgement to answer the question) is weak – meaning a lower mark:

Point: that much of London's population growth is due to immigration.

Evidence: gives India as an example of a country where many migrants have come from; also gives data from Figure 1 – that over 37% of the people living in Newham are Asian Indian British residents

Explanation: explains that migrants have come to the UK for jobs such as in construction and financial services

Point: that migrants have changed the character of parts of the city where they live

Evidence: for example, festivals like the Notting Hill Carnival in London

Explanation: explains how the shops or places of worship reflect the culture of migrants; names Wembley as an example

To get a top Level 3 mark, the candidate needs to make a stronger judgement. They state that immigration has had a big effect on cities but they need to expand on this more

Sample answer 2 This answer gets 3 marks and is a Level 1 answer because of the lack of evidence and detailed explanation in their points. For example:

Point: their point about London growing fast is general and not specifically linked to international migration

Explanation: they do explain this point – they state that London is growing fast because of immigrants from other countries; they explain how jobs attract people to London but don't give any examples

Point: that new restaurants and festivals open – referring to the changing character of the city because of migration

Explanation: the families of migrants come and join them, continuing to add to the population

However, the candidate does not make a judgement and the evidence / examples given are very broad and without any detail

Page 66 1 Plan your answer

Refer back to the guidance on page 35 for how to BUG the question.

Page 68 3 Write your answer

Answer should include:

- specific reference to your geographical investigation – what you did and where
- three points about the accuracy and reliability of your results – how did you try to ensure they were as reliable and accurate as possible?
- a clear understanding of the difference between accuracy and reliability
- a judgement about how accurate and reliable your results were; this should be a mini-conclusion at the end of your answer – a sentence or two which draws together your argument

Page 70 5 Mark different answers

Sample answer 1 This candidate achieves 8 marks and a top Level 3 answer because:

- they clearly explain how they ensured their results were accurate, for example, by using the BGS phone app and the flood risk maps from the Environment Agency
- they expand on this detail to explain what each data source showed them
- they recognise the difference between reliability and accuracy and use each term correctly
- they make a judgement at the end of each point and then again at the end of their answer – this ensures that they link back to the question

Sample answer 2 This candidate achieves 8 marks and a top Level 3 answer because:

- they make their judgements clear throughout – that their results were accurate in some ways but not others
- they clearly explain how they tried to ensure their results were as reliable as possible, for example, by using the IMD data and a decibel app on their phone
- they expand on this detail to explain what each data source showed them
- they recognise the difference between reliability and accuracy and use each term correctly

Page 75 3 Write your answer

Answer should include:

- three pieces of evidence for global climate change
- examples that support each piece of evidence for climate change – for example, some data about rising temperatures or rising sea levels
- a judgement about the strength of each piece of evidence (to ensure you evaluate)
- a mini-conclusion (a sentence or two) at the end that draws together your argument

Page 75 5 Mark different answers

Sample answer 1 This candidate achieves a low Level 2 and 4 marks. This is because although they've tried to use a PEEL structure in their paragraphs, they haven't evaluated. They haven't talked about how important each piece of data is.

To achieve a higher mark, they should remove one of their points and ensure they add the evaluation statements throughout. For example:

Point: one piece of evidence for climate change is that temperatures have risen globally since the nineteenth century

Evidence: records show that they have risen by about 0.8°C since the nineteenth century

Explanation: this is probably due to carbon emissions of greenhouse gases like CO_2 from burning fossil fuels

Evaluation / Link: it is hard to know exactly what temperatures were like in the nineteenth century because the equipment was less advanced. However, in the twenty-first century, equipment is incredibly accurate and shows a clear increase in global temperatures meaning that there is strong evidence for climate change from rising global temperatures

Sample answer 2 This candidate achieves 8 marks. This is because:

- they make three clear points; each point names a different piece of evidence for global climate change
- each point is explained and then evaluated; the candidate talks about whether the evidence is reliable or not and why
- by adding an evaluation statement to the end of each paragraph, they ensure that they are building an argument throughout and will achieve top marks for AO3

Page 81 1 Plan your answer

Refer back to the guidance on page 35 for how to BUG the question. Page 73 explains specifically how to do this for 'evaluate' questions.

Page 83 Write your answer

Answer should include:

- three points about how sediment deposition creates coastal landscapes
- examples / evidence that help expand your points – this evidence should be from Figure 1, explaining what is shows and your own knowledge
- a judgement about the strength of each point (to ensure you evaluate) – for example, if a coastal spit is the most significant

feature along a stretch of coast then sediment deposition does play a significant part in creating coastal landscapes

- a mini-conclusion (a sentence or two) at the end that draws together your argument and makes it clear what your final judgement is

Page 85 5 Mark a different answer

This answer gets 6 marks and is a Level 2 answer because it follows a PEEL structure. For example, in the first paragraph:

Point: the photo shows a spit that has formed through deposition

Explanation: the candidate explains how a spit forms through the process of longshore drift

Evidence: they refer back to the photo and talk about this specific spit

Link: they state how important deposition must be if it can divert river flow – linking back to the question and discussing the strength of the point (evaluation)

The candidate does this twice more to ensure they give three points. To get a higher mark the candidate should refer to the photo more often – for example in the second and third paragraphs. To ensure their judgement and evaluation is clear, they should also add a mini-conclusion at the end which draws together their argument

Key Geographical terms with definitions

*cross reference

A

abiotic non-living part of a *biome, includes the *atmosphere, water, rock and soil

abrasion the scratching and scraping of a river bed and banks by the stones and sand in the river

aftershocks follow an earthquake as the fault 'settles' into its new position

alluvium all deposits laid down by rivers, especially in times of flood

altitudinal zonation is the change in *ecosystems at different altitudes, caused by alterations in temperature, precipitation, sunlight and soil type

antecedent rainfall the amount of moisture already in the ground before a rainstorm

asthenosphere part of the Earth's *mantle. It is a hot, semi-molten layer that lies beneath the *tectonic plates

atmosphere the layer of gases above the Earth's surface

attrition the wearing away of particles of debris by the action of other particles, such as river or beach pebbles

B

bankful the *discharge or contents of the river which is just contained within its banks. This is when the speed, or *velocity, of the river is at its greatest

bar an accumulation of *sediment that grows across the mouth of a bay, caused by longshore drift

basalt a dark-coloured volcanic rock. Molten basalt spreads rapidly and is widespread. About 70% of the Earth's surface is covered in basalt *lava flows

biodiversity means the number of different plant and animal species in an area

biofuels any kind of fuel made from living things, or from the waste they produce

biogas a gas produced by the breakdown of organic matter, such as manure or sewage, in the absence of oxygen. It can be used as a *biofuel

biome a large-scale *ecosystem, e.g. tropical rainforest

biosphere the living layer of Earth between the *lithosphere and *atmosphere

biotic living part of a *biome, made up of plant (flora) and animal (fauna) life

black gold a term used for oil, as it is regarded as such a valuable commodity

bottom-up development experts work with communities to identify their needs, offer assistance and let people have more control over their lives, often run by *non-governmental organisations

brownfield sites former industrial areas that have been developed before

C

carbon dating uses radioactive testing to find the age of rocks which contained living material

carbon footprint a calculation of the total *greenhouse gas emissions caused by a person, a country, an organisation, event or product

carbon sequestration removing carbon dioxide from the atmosphere and locking it up in biotic material

carbon sinks natural stores for carbon-containing chemical compounds, like carbon dioxide or methane

Central Business District (CBD) the heart of an urban area, often containing a high percentage of shops and offices

channel refers to the bed and banks of the river

climatologist a scientist who is an expert in climate and climate change

collision zone where two *tectonic plates collide – forming mountains like the Himalayas

communism system of government, based on Karl Marx's theories; it believes in sharing wealth between all people

concordant coasts follow the ridges and valleys of the land, so the rock *strata is parallel to the coastline

connectivity how easy it is to travel or connect with other places

conservation means protecting threatened *biomes, e.g. setting up national parks or banning trade in endangered species

conservative boundary where two *tectonic plates slide past each other

constructive waves build beaches by pushing sand and pebbles further up the beach

continental crust the part of the Earth's crust that makes up land, on average 30-50 km thick

conurbation a continuous urban or built-up area, formed by merging towns or cities

convection currents transfer heat from one part of a liquid or gas to another. In the Earth's *mantle, the currents which rise from the Earth's core are strong enough to move the *tectonic plates on the Earth's surface

convergence a) the meeting of *tectonic plates; b) when air streams flow to meet each other

Coriolis force a strong force created by the Earth's rotation. It can cause storms, including hurricanes

cost-benefit analysis looking at all the costs of a project, social and environmental as well as financial, and deciding whether it is worth going ahead

counter-urbanisation when people leave towns and cities to live in the countryside

Covid-19 a highly contagious respiratory disease, first detected in 2019, that can lead to fatalities; in 2020 it spread rapidly around the world (as a *pandemic)

D

decentralisation shift of shopping activity and employment away from the *Central Business District (CBD)

deforestation the deliberate cutting down of forests to exploit forest resources (timber, land or minerals)

deindustrialisation decreased activity in manufacturing and closure of industries, leading to unemployment

delta a low-lying area at the mouth of a river where a river deposits so much *sediment it extends beyond the coastline

depopulation decline of total population of an area

deprivation lack of wealth and services. It usually means low standards of living caused by low income, poor health, and low educational qualifications

dip slope a gentle slope following the angle of rock *strata, found behind *escarpments

discharge the volume of water flowing in a river, measured in cubic metres per second

discordant coast alternates between bands of hard rocks and soft rocks, so the rock *strata is at right angles to the coast

dissipate means to reduce wave energy, which is absorbed as waves pass through, or over, sea defences

divergent plate boundary where two *tectonic plates are moving away from each other

diversification when a business (e.g. a farm) decides to sell other products or services in order to survive or grow

E

ecological debt when Earth's resources are being used up faster than Earth can replace them

ecological footprint is a calculation measured in global hectares (gha). It's the amount of land and water required to produce resources and deal with waste from each country

economic liberalisation when a country's economy is given the freedom of a 'market economy', consumers and companies decide what people buy based on demand

ecosystem a localized *biome made up of living things and their non-living environment. For example a pond, a forest, a desert

ecosystem services a collective term for all of the ways humans benefit from ecosystems

emerging economies countries that have recently industrialised and are progressing towards an increased role in the world economy

energy diversification getting energy from a variety of different sources to increase *energy security

energy security having access to reliable and affordable sources of energy

enquiry the process of investigation to find an answer to a question

epicentre the point on the ground directly above the focus (centre) of an earthquake

erosion means wearing away the landscape

escarpment a continuous line of steep slopes above a gentle *dip slope, caused by the erosion of alternate *strata

evacuate when people move from a place of danger to a safer place

evaporation the changing of a liquid into vapour or gas. Some rainfall is evaporated into water vapour by the heat of the sun

F

fault large cracks caused by past tectonic movements

fetch the length of water over which the wind has blown, affecting the size and strength of waves

fieldwork means work carried out in the outdoors

flood plain flat land around a river that gets flooded when the river overflows

focus the point of origin of an earthquake

food miles the distance food travels from the producer to the consumer. The greater the distance, the more carbon dioxide is produced by the journey

food web a complex network of overlapping food chains that connect plants and animals in *biomes

formal economy means one which is official, meets legal standards for accounts, taxes, and workers' pay and conditions

fossil fuels a natural fuel found underground, buried within sedimentary rock in the form of coal, oil or natural gas

free trade the free flow of *goods and *services, without the restriction of tariffs

friction the force which resists the movement of one surface over another

G

gentrification high-income earners move into run-down areas to be closer to their workplace, often resulting in the rehabilitation and *regeneration of the area to conform with middle class lifestyles

geographical conflict means disagreement and differences of opinion linked to the use of places and resources

geographical information systems (GIS) a form of electronic mapping that builds up maps layer by layer

geothermal heat from inside the Earth

glacial a cold period of time during which the Earth's glaciers expanded widely

global circulation model a theory that explains how the *atmosphere operates in a series of three cells each side of the Equator

global shift change in location of where manufactured goods are made, often from developed to developing countries

globalisation increased connections between countries

goods physical materials or products that are of value to us

green belt undeveloped areas of land around the edge of cities with strict planning controls

greenhouse effect the way that gases in the atmosphere trap heat from the sun. Like the glass in a greenhouse – they let heat in, but prevent most of it from escaping

greenhouse gases gases like carbon dioxide and methane that trap heat around the Earth, leading to global warming

gross domestic product (GDP) the total value of *goods and *services produced by a country in one year

groundwater flow movement of water through rocks in the ground

H

hard engineering building physical structures to deal with natural hazards, such as sea walls to stop waves

helicoidal flow a continuous corkscrew motion of water as it flows along a river channel

holistic management takes into account all social, economic and environmental costs and benefits. In coastal management this means looking at the coastline as a whole instead of an individual bay or beach

hot spot columns of heat in Earth's *mantle found in the middle of a tectonic plate

Human Development Index (HDI) a standard means of measuring human development

hydraulic action the force of water along the coast, or within a stream or river

hydrological cycle the movement of water between its different forms; gas (water vapour), liquid and solid (ice) forms. It is also known as the water cycle

hyper-urbanisation rapid growth of urban areas

I

Index of Multiple Deprivation (IMD) means of showing how deprived some areas are

indigenous peoples are the original people of a region. Some indigenous groups still lead traditional lifestyles, e.g. a tribal system, hunting for food

industrialisation where a mainly agricultural society changes and begins to depend on manufacturing industries instead

infiltration the soaking of rainwater into the ground

informal economy means an unofficial economy, where no records are kept. People in the informal economy have no contracts or employment rights

infrastructure the basic services needed for an industrial country to operate e.g. roads, railways, power and water supplies, waste disposal, schools, hospitals, telephones and communication services

interception zone the capture of rainwater by leaves and branches. Some *evaporates again and the rest drips from the leaves to the soil

interglacial a long period of warmer conditions between *glacials

interlocking spurs hills that stick out on alternate sides of a V-shaped valley, like the teeth of a zip

intermediate technology uses low-tech solutions using local materials, labour and expertise to solve problems

Inter-Tropical Convergence Zone (ITCZ) a narrow zone of low pressure near the Equator where northern and southern air masses converge

invasive species (or alien species) is a plant, animal or disease introduced from one area to another which causes ecosystem damage

irrigation is the artificial watering of land that allows farming to take place

J

jet streams high level winds at around 6-10km that blow across the Atlantic towards the UK

joints small and usually vertical cracks found in many rocks

L

lagoon a bay totally or partially enclosed by a *spit, *bar or reef running across its entrance

landslide a rapid *mass movement of rock fragments and soil under the influence of gravity

latitude how far north or south a location on the Earth's surface is from the Equator, measured in degrees

lava melted rock that erupts from a volcano

lava flows *lava flows at different speeds, depending on what it is made of. Lava flows are normally very slow and not hazardous but, when mixed with water, can flow very fast and be dangerous

level of development means a country's wealth (measured by its GDP), and its social and political progress (e.g. its education, health care or democratic process in which everyone can vote freely)

lithosphere the uppermost layer of the Earth. It is cool and brittle. It includes the very top of the *mantle and, above this, the crust

M

magma melted rock below the Earth's surface. When it reaches the surface it is called *lava

magnitude of an earthquake (how much the ground shakes), an expression of the total energy released

mantle the middle layer of the Earth. It lies between the crust and the core and is about 2900 km thick. Its outer layer is the *asthenosphere. Below the asthenosphere it consists mainly of solid rock

mass movement the movement of material downslope, such as rock falls, *landslides or cliff collapse

megacity a many centered, multi-city urban area of more than 10 million people. A megacity is sometimes formed from several cities merging together

middle course the journey of a river from its source in hills or mountains to mouth is sometimes called the course of the river. The course of a river can be divided into three main sections a) upper course b) middle course and c) lower course

migration movement of people from one place to another

Milankovitch cycles the three long-term cycles in the Earth's orbit around the sun. Milankovitch's theory is that *glacials happen when the three cycles match up in a certain way

mudflats flat coastal areas formed when mud is deposited by rivers and coasts

multicultural a variety of different cultures or ethnic groups within a society

multiplier effect when people or businesses move to an area and invest money on housing and services, which in turn creates more jobs and attracts more people

N

natural increase the birth rate minus the death rate for a place. It is normally given as a % of the total population

natural resources are materials found in the environment that are used by humans, including land, water, fossil fuels, rocks and minerals and biological resources like timber and fish

net primary productivity (NPP) a measure of how much new plant and animal growth is added to a biome each year

non-governmental organisation (NGO) NGOs work to make life better, especially for the poor. Oxfam, the Red Cross and Greenpeace are all NGOs

non-renewable energy sources that are finite and will eventually run out, such as oil and gas

northern powerhouse a major core region of cities (with a similar population to London) that has the potential to drive the economy of northern England

Key Geographical terms with definitions

nutrient cycle nutrients move between the biomass, litter and soil as part of a continuous cycle which keeps both plants and soil healthy

O

ocean currents permanent or semi-permanent large-scale horizontal movements of the ocean waters

oceanic crust the part of the Earth's crust which is under the oceans, usually 6-8 km thick

Organisation of Petroleum Exporting Countries (OPEC) established to regulate the global oil market, stabilize prices and ensure a fair return for its 12 member states who supply 40% of the world's oil

outsourcing using people in other countries to provide services if they can do so more cheaply e.g. call centres

ox-bow lake a lake formed when a loop in a river is cut off by floods

P

pandemic disease which spreads to (almost) every country

Pangea a supercontinent consisting of the whole land area of the globe before it was split up by continental drift

peak oil the theoretical point at which half of the known reserves of oil in the world have been used

plate boundaries where *tectonic plates meet. There are three kinds of boundary a) *divergent – when two plates move apart b) *convergent – when two plates collide c) conservative – when two plates slide past one another

plumes upwelling of molten rock through the *asthenosphere to the *lithosphere

plunging waves typically tall and close together, created by strong winds

population density the average number of people in a given area, expressed as people per km²

population structure the number of each sex in each age group usually displayed in a population pyramid diagram

poverty line the minimum level of income required to meet a person's basic needs (US$1.90)

predict saying that something will happen in the future. A scientific prediction is based on statistical evidence

prevailing winds the most frequent direction the wind blows in a certain area

primary effects the direct impacts of an event, usually occurring instantly

primary products raw materials

Purchasing Power Parity (PPP) shows what you can buy in each country, now used to measure *GDP

pyroclasts fragments of volcanic material that is thrown out during explosive eruptions

Q

quality of life a measure of how 'wealthy' people are, but measured using criteria such as housing, employment and environmental factors, rather than income

Quaternary the last 2.6 million years, during which there have been many *glacials

R

radioactive decay atoms of unstable elements release particles from their nuclei and give off heat

rebranded a change of image

recurved hooked

regeneration means re-developing former industrial areas or housing to improve them

renewable a resource that does not run out and can be restored, such as wind or solar

re-urbanisation when people who used to live in the city and then moved out to the country or to a suburb, move back to live in the city

Richter scale a scale for measuring the magnitude of earthquakes

rock outcrop a large mass of rock that stands above the surface of the ground

rockfalls a form of *mass movement where fragments of rock fall freely from a cliff face

rural-urban fringe the area where a town or city meets the countryside

rural-urban migration the movement of people from the countryside to the cities, normally to escape from poverty and to search for work

S

Saffir-Simpson Hurricane Scale a scale that classifies hurricanes into five different categories according to their wind strength

salt marsh salt-tolerant vegetation growing on mud flats in bays or estuaries. These plants trap sediments which gradually raise the height of the marsh. Eventually it becomes part of the coast land

saltation the bouncing of material from and along a river bed or a land surface

sand dune onshore winds blow sand inland, forming a hill or ridge of sand parallel to the shoreline

saturated soil is saturated when the water table has come to the surface. The water then flows overland

scree angular rock pieces created by freeze-thaw weathering

secondary effects the indirect impacts of an event, usually occurring in the hours, weeks, months or years after the event

secondary products manufactured goods

sediment material such as sand or clay that is transported by rivers

seismometer a machine for recording and measuring an earthquake using the *Richter scale

services functions that satisfy our needs

shore-line platform the flat rocky area left behind when waves erode a cliff away

Shoreline Management Plan (SMP) this is an approach which builds on knowledge of the coastal environment and takes account of the wide range of public interest to avoid piecemeal attempts to protect one area at the expense of another

slope processes cause *mass movement or *soil creep

soft engineering involves adapting to natural hazards and working with nature to limit damage

soil creep the slow gradual movement downslope of soil, *scree or glacier ice

solution chemicals dissolved in water, invisible to the eye

spatial means 'relating to space' e.g. the spatial growth of a city means how much extra space it takes up as it grows

spit a ridge of sand running away from the coast, usually with a curved seaward end

stakeholders a person with an interest or concern in something, such as those who are likely to be affected by natural hazards

storm hydrograph a graph which shows the change in both rainfall and discharge from a river following a storm

storm surge a rapid rise in the level of the sea caused by low pressure and strong winds

strata distinctive layers of rock

stratosphere the layer of air 10-50km above the Earth's surface. It is above the cloudy layer we live in, the troposphere

strip mining (or open-pit, opencast or surface mining) involves digging large holes in the ground to extract ores and minerals that are close to the surface

studentification communities benefit from local universities which provide employment opportunities and a large student population which can regenerate pubs, shops and buy-to-let properties

sub-aerial processes occurring on land, at the Earth's surface, as opposed to underwater or underground

subduction describes *oceanic crust sinking into the *mantle at a convergent *plate boundary. As the crust subducts, it melts back into the mantle

subsistence farming where farmers grow food to feed their families, rather than to sell

suburbanisation the movement of people from the inner suburbs to the outer suburbs

surface run-off rainwater that runs across the surface of the ground and drains into the river

suspension tiny particles of *sediment dispersed in water

Sustainable Development defined by the Brundtland Commission as that which 'meets the needs of the present without compromising the ability of future generations to meet their own needs'

sustainable management meeting the needs of people now and in the future, and limiting harm to the environment

T

tar sands sediment that is mixed with oil, can be mined to extract oil to be used as fuel

tectonic hazards natural events caused by movement of the Earth's plates that affect people and property

tectonic plate the Earth's surface is broken into large pieces, like a cracked eggshell. The pieces are called tectonic plates, or just plates

terms of trade means the value of a country's exports relative to that of its imports

thalweg the line of the fastest flow along the course of a river

thermal expansion as a result of heating, expansion occurs. When sea water warms up, it expands

throughflow the flow of rainwater sideways through the soil, towards the river

till sediment deposited by melting glaciers or ice sheets

top-down development when decision-making about the development of a place is done by governments or large companies

topography the shape and physical features of an area

traction force that rolls or drags large stones along a river bed

transnational companies (TNCs) those which operate across more than one country

transpire when plants lose water vapour, mainly through pores in their leaves

tsunami earthquakes beneath the sea bed generate huge waves that travel up to 900km/h

U

urbanisation means a rise in the percentage of people living in urban areas, compared to rural areas

V

velocity the speed of a river, measured in metres per second

Volcanic Explosivity Index (VEI) measures the explosiveness of volcanic eruptions on a scale of 1 to 8

W

water table the upper limit of saturated rock below the ground

weathering the physical, chemical or biological breakdown of solid rock by the action of weather (e.g. frost, rain) or plants

wildfire uncontrolled burning though forest, grassland or scrub. Such fires can 'jump' roads and rivers and travel at high speed

world cities trade and invest globally e.g. London and New York